Raw Flows

Edition Angewandte

Book Series
of the University of
Applied Arts Vienna

Gerald Bast,
Rector (Ed.)

edition: 'ʌngewʌndtə

Raw Flows

Fluid Mattering in Arts and Research

Roman Kirschner (Ed.)

DE GRUYTER

List of contents

Foreword

Roman Kirschner

Matter is in flux. Its flows can be encountered on different scales of space and time. The characteristics of these flows influence researchers in their active and direct material engagement. This publication investigates how fluidity and flow carve their specific paths into experimental practices and thinking patterns using examples from the arts, humanities and sciences.

The material property and general phenomenon of being fluid, is extraordinarily meaningful, as it represents a basic and irreducible principle of processuality, change and open-ended becoming. In many respects it is related to and involved in matter's generative aspects by bringing together substances, mixing them or helping them to react and transform.[1] Furthermore, it also supports catabolic tendencies in the sense that it facilitates the breakdown and decay of forms and the increase of entropy.[2] Thus, fluidity reminds us on the one hand that the world is in constant movement and on the other hand that this movement is irreversible. And although many aspects of becoming are subject to chance, and could therefore be seen in a fatalistic light, the world is nevertheless excitable and not just stuck in an eternal equilibrium.[3]

The French writer Francis Ponge gave an impressive account of two main aspects of fluidity in his short text on water, the most prominent and abundant fluid. First, he points out that gravity is its most efficient accomplice and vice, writing, "Forever lower: that seems to be its motto."[4] And on its travel downwards, so excellently facilitated by its formlessness, water circumvents, perforates, infiltrates and erodes.[5] And second, he states how it always escapes him and slips through his fingers. "[W]ater eludes me, eludes all definition, yet leaves its traces in my mind […]."[6] Gravity's pull on a flexible mass and its conceptual elusiveness as a result of its constant rebellion against containment and fixation characterize fluidity's appearance.

The existential impact of all of these aspects and fluids' experiential richness can explain why fluidity and its mostly overlapping companion liquidity have become widely used metaphors that inextricably oscillate between fascination and thread.[7] But the challenge of *Raw Flows* is to deal with fluidity beyond its metaphorical use. What therefore seems crucial is to understand fluidity as a phenomenon that is first of all deeply rooted in the material world. It is from there that it draws its influence, pervasive power and heterogeneity in appearance.

This publication intends to enrich the experience and understanding of fluid phenomena, while not denying or omitting the importance of its metaphorical impact and scope. To achieve this goal, the authors concentrate on fluidity's material basis and wider materiality – in the sense of the network of material relations that the stuff that brings forth or shows fluidity, unfolds or draws from. On a spectrum reaching from concrete substances through the manifold and heterogeneous entanglements of materiality to metaphors involving fluidity, metaphors are clearly at the opposite pole of stuff. Due to their derivative nature, metaphors can help verbalize aspects and parallels that would otherwise perhaps be inexpressible, yet they hardly reveal details about the actual phenomenon of fluidity itself. As for the more illuminating network of relations that materiality stands for, the authors of *Raw Flows* avoid using abstract notions of matter in order to ground it and rather start all investigations and descriptions from actual substances or stuff[8] wherever this was possible.

In this book, dealing with fluidity does not necessarily require dealing with wetness. Because not only liquids like the aforementioned water but also all kinds of amorphous materials, which are rheologically situated in between the aggregate states of solid and liquid, can show fluid behaviour. That is why polymers and even granular materials like diamond dust are featured in *Raw Flows'* contributions next to lubricants, watery solutions, blood, or photonic crystals in oily suspensions. The different flows of these materials, are being followed by the authors through different scales wherever they appear, for example as advection or shearing at the macro level, as the interplay between microstructures at the mesoscale, or as fluctuations and self-propulsions in even smaller areas. Thus, the presented experiences, ideas, concepts and drawn conclusions can stay as close as possible to actual phenomena and practices.

But as mentioned before, substances and stuff are also integrated in and inseparable from contexts and systems, be it a material milieu, a research environment, or a wider cultural/historical landscape. In their physical, chemical, biological and cultural embeddedness they are active or acted upon. In this setting, but more precisely as a result of both levels, the material basis and the wider context of materiality, fluidity appears and can be modified, adjusted, reconfigured in diverse ways. Furthermore, it is also an effect of both levels that fluidity allegedly contributes to the formation of knowledge structures and meaning.

The paradox of polymorphosis

In Georges Didi-Huberman's text, *The Order of Material: Plasticities, Malaises, Survival*,[9] wax is the protagonist of his inquiries which reach from a phenomenological approach to art historical nexuses and psychoanalytical interpretations. There, in accordance with the general approach used in this book, Didi-Huberman bemoans that materials always come second after form or mind in art history. Referring to Sartre's famous passage on viscosity in *Being and Nothingness*, where he describes material quality as *revelatory of being*,[10] Didi-Huberman sets out to weave a network of meanings around the viscosity of wax. He ascribes to viscosity a kind of activity and intrinsic power, which he calls the power of metamorphosis or polymorphosis.[11] This power draws from the paradox of wax's special position between being solid and liquid, as it can pass from one state to the other without much effort and within a small range of change in temperature and pressure. This ease of changing states facilitates its oscillation between incorporating a form and being formless and bridges the abstract contradiction between form and formlessness. Similarly, some materials appearing in this book play with this border between form and formlessness in their own way. They can, for example, keep their form only under sudden high pressure while the rest of the time, they are subject to a very slow flow, which tends to flatten them out. They are called non-Newtonian fluids, like Polyvinyl alcohol-based polymers or the silicone-based Silly Putty. Like liquids they have some short-range order at the atomic length scale but no long-range order that leads to this flow.[12] Not only by them but also by all of the fluid materials featured in this book, the traditional static understanding of materials gets categorically undermined. Their fluidity and fluidity in general cannot be understood in static terms, although snapshots or freeze frames can of course help in the approach or in understanding specific situations. Unfortunately though, still images fail to communicate the core feature, namely processuality, which can be further described by movement, continuity and the rich simultaneity and parallelism of happenings. The material interactions and propagating impulses on fluids from the outside can and do often lead to emerging patterns and forms like streams, waves, drops, splashes and turbulences. And these can vary significantly between different fluid substances–which characterizes them and makes them distinguishable. That is why the chemist and philosopher Jens Söntgen does not agree with the ascription of formlessness to liquids. He rather subsumes them, together with all other substances, under the term *Gebilde*,[13] a German term meaning that something has gone through a specific process of formation. And it is due to this process of formation that all stuff shows its very own and characteristic structure, its Eigenformen, which furthermore distinguishes it from the reductionist notion of ideal substances or structural formulae that presume uniform aggregates of atoms and molecules. However, the forms of liquids, and this is also true for fluids, are ephemeral and not as long-lasting as those of solids. Their polymorphism, their back and forth movement between forms, happens at a more or less high speed, depending on their viscosity. The material properties and massive material interactions within fluid and even more so within liquid substances are the reason why it is so hard for the sciences to create a mathematical model that is not just a *perfect fluid*, which means that fluids are stripped of all of their peculiarities by omitting shear stresses, heat conduction, and viscosity. Thus, by not so much showing the current limits of computability but of formalization, fluidity undermines approaches of formal description. Beyond simulation and modeling, it is equally challenging to repeat the creation of the exact same forms and constellations in fluids given the amount of parameters that influence their formation in real space and time.

On the level of materiality, where conceptual and cultural influences interfere, the elusive phenomenon of fluidity is as difficult to grasp as on the formal and material levels. To cope with this fact, this book was initially thought of as a vessel in which the processuality of fluids might temporarily and carefully be trapped with the hope of not excluding and losing too much through its short-term containment. But the idea of a vessel proved to be inappropriate from the beginning as the initial concept of *Raw Flows* found itself in the middle of surprising outspills and overflows – some of the most common accidents that influence researchers in their everyday handling of flows. The number of examples was so abundant, their diversity so large, yet the vessel's ineptness so obvious on such different scales that the whole mass became untenable and overshot while its remaining flows got blocked.

Consequently, the most adequate approach beyond containment proved to be the use of an irregular and cross-disciplinary grid immersed in the material flows of the world at some selected locations with which to catch, extract, and evaluate ways of dealing with fluidity. This way it was possible to get insights into specialized, self-conscious practices which in their turn provided the backdrop for further exploration and discussion of the roles that fluidity and flow may play in relation to the epistemic, aesthetic and experiential. The grid's cross-disciplinarity was caused by the fact that, from the beginning, the focus was set on fluidity as a material property and phenomenon and not on some disciplinary homogeneity. The general endeavor started in the arts and especially in the art-based research of the project *Liquid Things* with its epistemic interest derived from art production within the field of plastic arts. But due to the wide spectrum of fluidity, it branched out into a cross-disciplinary effort and now integrates contributions from disciplines as diverse as history of science, art history, fluid dynamics, design and cultural studies. The irregularity of the grid resulted from the aforementioned need to track fluidity through different materials, different scales, and different levels of materiality in order to capture its inherent dynamics and related research practices.

The contributions

Most of the contributions to the book use contemporary or recent examples but they differ largely in their perspectives and scopes. The science historian Hans-Jörg Rheinberger opens the topic up wide by discussing liquid metaphors in scientific research including dynamics like meandering, turbulence, or influences, and behaviors facing obstacles. Subsequently, he concentrates more closely on large historical apparatuses from the early life sciences which he contrasts with the tiny amount of liquids they operate with and how. Additionally, he examines different boundary operations between soft/hard, wet/dry, and living/non-living in relation to specific research procedures.

From the perspective of cultural studies, Benjamin Steininger extends the issue of the interplay between machines and fluids by focusing on lubricants as transmitters of energy that also facilitate the kinetic closure of machines. He discusses the materiality, mediality, and agency of lubricants in the historical case study of Walter Oswald's search for optimal lubricity. As a curious fact, Oswald achieved the task of minimizing friction through a protective oily layer of active molecules that counter intuitively attack the surface of the adjacent metals they are supposed to preserve.

The art historian Inge Hinterwaldner investigates different roles which fluid or granular materials take in the studies of turbulence phenomena and flow analysis. Her thorough examination of different experimental setups brings her to the insight that the materials used as substrates, media or markers were not just protagonists or extras. Instead, their specific characteristics in the way they transport information turned them into objects of study in their own right.

The contribution of physicist and hydrodynamics specialist, Jean-Marc Chomaz, takes the reader from Hinterwaldner's room-sized experimental setups to much larger scales. He starts from the idea of a *universal flow*, which he illustrates by discussing material flows of particles, photons and neutrinos through to the dynamic system of the universe. He thereby specifies phenomena such as structural formation processes of proto-stars or the cloud dynamics in the earth's atmosphere. Closer to our planet, he discusses exchanges between oceans and the atmosphere which also leads him to outlining the collaborative project, *Luminiferous Drift* developed together with the artist couple Evelina Domnitch and Dmitry Gelfand.

Evelina Domnitch and Dmitry Gelfand follow with a chapter about their project *Photonic Wind* on the fluidity of light, which can be evoked through a phenomenon known as photophoresis, light-induced levitation and migration of matter. Starting from earlier works on acoustic levitation they delve into the development process and the material experiences they encountered during their intricate work with diamond dust and photons in the nearly complete void of vacuum chambers. The outcome of their experiments for *Photonic Wind* were shown in the exhibition *Kontinuum* to which the following contributions also refer either directly or indirectly by describing preceding work processes, experiments and reflections.

Thus, with a few impressions from the Viennese gallery *Im Ersten*, in which Kontinuum was shown in early 2015, some glimpses on interim results of the research project *Liquid Things* are provided. These visual clues to the actual material experiences of the developed materials and their specific integrations into spatial installations can help readers to imagine the possible directions of impact in the following texts.

The artist and researcher, Roman Kirschner, investigates parallels, intersections and mutual catalyses between the dynamics of matter and imagination. Starting from an ecological approach into material and mental entanglements, he undertakes a preliminary exploration of material signs and investigates the outlines of a contemporary, non-romantic material imagination in a larger framework of matter/energy flows.

Subsequently, the art historian Esther Moñivas Mayor takes a close look at the project studio for *Liquid Things*, documenting practices and processes from a systemic yet experimental perspective. Moñivas's essay, which is based on her detailed, persistent observations and personal experiences during her collaboration period of more than six months, is guided by material properties. Her meandering narration reconstructs the interactions between (liquid) things, (diverse) people, (an ever changing) environment and (fluid) imaginations.

In the last contribution, which concludes the book, the architect and design researcher, Karmen Franinović, discusses research methods and techniques facing fluid and active materials. She examines the trans-disciplinary workshop *Material Aktiv Denken* to develop an understanding of the dynamic interactions between materials, environments and people, which resulted from probing different workshop conditions and theory-practice techniques merging material and conceptual processes. Finally, she proposes the notion of *fluid affordance*, which opens up to flows of materials and thoughts that escape the solidification of things and concepts.

The collected inquiries and arising questions aim at influencing ongoing discussions in art-based research and in the wider field of New Materialisms. Furthermore, this publication wants to put some of these arising issues into productive friction: the implications of thinking with and conceiving of continua instead of operating with parts/particles; knowledge structures in the making and in constant transformation; materials as qualitative and non-representational entities outside the framework of the expected; the activities of fluid materials themselves and within research processes; etc. The overall goal is to provide grounded insights into the complex phenomenon of fluidity to discuss exemplary historical and contemporary ways of dealing with fluidity and to set up an informed basis for contemporary art discourse and practice that can also feed back into other disciplines.

1 Earley, Joseph E. Sr.: *Modes of Chemical Becoming*, in: *Hyle. International journal for philosophy of chemistry*, Vol. 4, No. 2, 1998, pp. 105-115, Retrieved Apr. 2, 2016, from http://www.hyle.org/journal/issues/4/earley.pdf

2 Arnheim, Rudolf: *Entropy and Art. An essay on disorder and order*, University of California Press 1971

3 See e.g.: Prigogine, I., Stengers, I.: *Dynamics from Leibniz to Lucretius*, in: Serres, Michel: *Hermes. Literature, science, philosophy*, Johns Hopkins University Press 1982, p. 135-158

4 Ponge, Francis: *Water,* in: *The nature of things*, Red Dust 1995, p. 28; French original: "Toujours plus bas : telle semble être sa devise […]" in: Ponge, Francis: *De l'eau,* in: *Le Parti pris des choses* (1942), Athlone Press 1979, p. 55

5 ibid., p. 55

6 Ponge: *Water*, p. 29; French original: „[E]lle m'échappe, échappe à toute définition, mais laisse dans mon esprit […] des traces […].

7 e.g. the concepts of *cognitive fluidity or liquid modernity* in: Mithen, S.: The prehistory of the mind, Thames and Hudson 1996 and Bauman, Z: Liquid Modernity, Polity Press 2000

8 The term *stuff* is used as in: Ruthenberg, K. et al. (eds.): *Stuff. The nature of chemical substance*, Königshausen und Neumann 2008

9 Didi-Huberman, Georges: *The Order of Material: Plasticities, malaises, survival* (Matlock, J, Trans.), in: Taylor, B, (ed.): *Sculpture And Psychoanalysis*. (195 - 212). Ashgate Publishing 2006

10 Sartre, Jean-Paul: *Being and Nothingness*, Routledge 1972, p. 600, cited after Didi-Huberman, p. 201

11 For another relevant perspective on viscosity in this context see DeLanda, Manuel: *Nonorganic Life*, in: Crary, J. et al. (eds.): *Incorporations (Zone 6)*, Zone 1992, p. 129-167

12 „Short range is defined as the first- or second-nearest neighbours of an atom. […] fluids, such as water, have short-range order but lack long-range order." (At longer distances, the positions of the atoms become uncorrelated.) from Mahan, Gerald D.: *Crystal* in: *Encyclopaedia Britannica*, Retrieved Apr 16, 2016, from http://www.britannica.com/science/crystal#ref506252

13 Söntgen, Jens: Stoffe und Dinge, p. 8-11, Retrieved March 23, 2016, from https://opus.bibliothek.uni-augsburg.de/opus4/files/1543/Soentgen_Stoffe_und_Dinge.pdf; in an english text on stuff, *Gebilde* is inadequately translated as *characteristic structures*. See Soentgen, J.: *Stuff: A phenomenological definition*, p. 80-82, in: Ruthenberg 2008

Acknowledgements

I would like to express my gratitude to all the contributing authors and to the following people for their support during different stages of the project Liquid Things:

Partners, collaborators, and supporters

Yunchul Kim, Karmen Franinović (Zurich University of the Arts), Silvain Michel (Swiss Federal Laboratories for Materials Science and Technology), Florian Wille (Zurich University of the Arts), Ruairi Glynn and Benjamin Haworth (Bartlett School of Architecture), Manuel Kretzer (ETH Zurich), Aernoudt Jacobs (Overtoon Brussels), Tobias Nöbauer (University of Vienna), Unyong Kim and Indra Jäger (Gallery Im Ersten), Constantin Luser, Philipp Haffner, Anton Ovidiu, Karin Haas, Heiko Schmid, Thomas Laureyssens, Matthias Tarasiewicz, Lucia Ayala, Georg Trogemann (Academy of Media Arts Cologne), Allison Kudla (Institute for Systems Biology Seattle), and Jens Hauser (University of Copenhagen)

Supporters at the University of Applied Arts Vienna

Institute of Fine Arts and Media Arts / Art and Science:
Virgil Widrich (Project Director of Liquid Things), Bernd Kräftner, Valerie Deifel, Juliana Herrero, Sonja Orman

Support Art and Research:
Alexander Damianisch, Angelika Zelisko, Wiebke Miljes, Franziska Echtinger

Institute of Art and Technology / Technical Chemistry and Science Visualization:
Alfred Vendl, Bernhard Pichler, Leonhard Gruber

Information, Publication and Event Management:
Anja Seipenbusch-Hufschmied

The research was funded by the Austrian Science Fund (FWF): AR 112-G21 in the Arts-based research programm PEEK

In Constant Flux: Thoughts about the Epistemic

Hans-Jörg Rheinberger

This text is the amended transcript of a talk given on October 11, 2013, during the symposium *Flows (Un)Bound* at the University for Applied Arts Vienna.

I would like to start my presentation with a few words about the nature of the environments in which experimental scientists work and the materials they become involved with in their working spaces. I think that, to a certain extent, correlations will show up between what scientists are doing in their laboratories and what artists do in their studios. There are, of course, differences as well. We should not ignore them but rather see how these two spaces can be brought into fruitful interaction. Let me recall the art historian George Kubler, who wrote, in 1962, a little book with the title *The Shape of Time. Remarks on the History of Things*.[1] In this book, Kubler tries to bind the history of the sciences and the history of the arts together, and characterizes the situation in which a productive scientist or artist finds him or herself, not with a metaphor of fluidity – I will come back to a number of these metaphors later – but with the metaphor of a mine. He describes the position of the exploring artist – or scientist for that matter – as standing at the end of a mine shaft that has been dug by others: On the one hand, the actual state is determined by what happened before; on the other hand, it is not clear and there is no signpost anywhere in this labyrinth that can tell him or her which direction to take from the end-point where he or she is actually standing. A contemporary of Kubler, historian of science Thomas Kuhn, whose famous book *The Structure of Scientific Revolutions* appeared that same year 1962,[2] once characterized the scientific enterprise as a process *driven from behind*, which means that it does not actually obey the usual teleological, goal-directed ideas about what to do next.[3] It is not oriented toward a final point omega, it rather tries to get away from the actual state of the art.

Let me come to the notion of *epistemic* that stands in the title of this talk. It is closely related to the notion of epistemic things. When I, as a historian of science, decided to look in more detail at what is going on in experimental environments, I came to settle on two interacting components of what scientists call their experimental systems: they cannot be conceived of without each other but must, nevertheless, be distinguished according to their specific function, respectively. One of them is the the object being worked on, including its unknown aspects – the latter makes me prefer to talk about things rather than objects. The notion of an object carries with it a certain kind of definiteness and determination that the notion of a thing is lacking. On the other hand, an experimental setup is characterized by a number of more or less specified procedures, apparatus, and instruments that can be identified as the technical conditions under which epistemic things take shape and are explored.

The idea that there is something at the frontiers of research that is not completely under control is an experience that many scientists have reported and written about over the past 200 years, that is, roughly the period in which scientific experimentation as we know it today took shape. Let me give just two examples from the nineteenth century. In one of his notebooks, the French physiologist Claude Bernard says: "C'est le vague, l'inconnu qui mène le monde,"[4] meaning that it is the vague, the nebulous that moves the world. By *le monde*, he is clearly addressing the world of science in which he lived. And in the memoirs of Bernard's contemporary, the German physiologist and physicist Hermann von Helmholtz, we find the following image:

> *I had to compare myself with a mountain climber who, without knowing his way, ascends slowly and arduously and who, sometimes by reflecting, sometimes by chance, detects new traces of paths that bring him another step further.*[5]

Liquidity as metaphor

We move now from metaphors related to mining and climbing an impassable mountain to another frequently-used metaphor in this context, one that is close to the topic of this conference, that of a stream of water. If you take a look at Ludwik Fleck's book on the *Genesis and Development of a Scientific Fact*,[6] you will find that he compares the research process to a meandering river. Of course, he says, we all know that gravity has something to do with the flux of water, but gravity by no means determines the concrete ways in which a river's bed is shaped by its own flow. As it goes along, it encounters obstacles; sometimes it has to flow around barriers, or carve breakthroughs, and other things of this sort. We have a whole little universe of images associated with the idea of a meandering river that is carving its own bed while flowing from the top of the mountain to the bottom of the sea. Just to give another example, there is the very interesting book – actually a collected volume – called *Eléments d'histoire des sciences* that came out in the late 1980s in France and was edited by Michel Serres.[7] In his introduction, Serres actually uses the same image of a meandering river to characterize the flux, the process of the development of the sciences, including backwaters and impasses where the water comes to a standstill or even has to reverse its flow. I will return to this. I think, if we are talking about fluxes and flows, we also have to talk about the stops that punctuate this movement.

The historian of science Yehuda Elkana, while looking at a particular episode in the history of physics of the nineteenth century, namely the development of the concept of force in the writings of the already mentioned Herman von Helmholtz, explicitly talks about a *concept in flux* in this respect.[8] Here, the agency lies less in a concrete experimental trajectory and the associated pathways of investigation,[9] but more in the conceptual changes that are associated with scientific work.

That is probably enough for drawing attention to a few aspects of the – fundamentally non-teleological – dialectics between epistemicity and technicity in experimental trajectories, as one could put it in summing up. What I would like to do now is to go through an example of an epistemic thing in flux that has massively shaped the life sciences of the twentieth century: I talk about what came to be called the *gene*.[10] Around the beginning of the twentieth century, the notion of the gene took root along with the emergence of heredity as a scientific discipline in its own right that was to be baptized genetics accordingly. Genetics continues to play an important role in the life sciences as well as in our everyday life today, and in an ever-increasing way even. At the beginning of the twentieth century, the gene looked considerably more abstract than today. People were trying to identify something that they called a factor. Indeed, several other terms were coined before William Bateson picked up a notion that Hugo de Vries had already been using before and termed the entity a gene. It was not visible at the time, nobody could physically touch it, but it nevertheless seemed to determine the appearance of organisms. It was conceived of as something like an atom of life.

Let us look at a sketch taken from the notes of one of the geneticists of the early twentieth century, Carl Correns (Fig. 1) whose notebooks are preserved in the Archive of the Max Planck Society in Berlin. Here, we can see how Correns tried to make sense of what he himself did not yet call a gene – but an *Anlage*. It is a very schematic way of making sense of his experiments. The principle is that the *Anlagen* always consist of two halves that belong to each other but that can be separated independently along the sequence of Anlagen during reproduction. Actually, the experimental practice behind this kind of reasoning – the hybridization of organisms – did not allow Correns and his contemporaries to go into more detail. However, by crossing different organisms with each other and looking at the visible characteristics of the progeny, they concluded from the regularities of the distribution pattern that these characteristics referred to something hidden deep in the

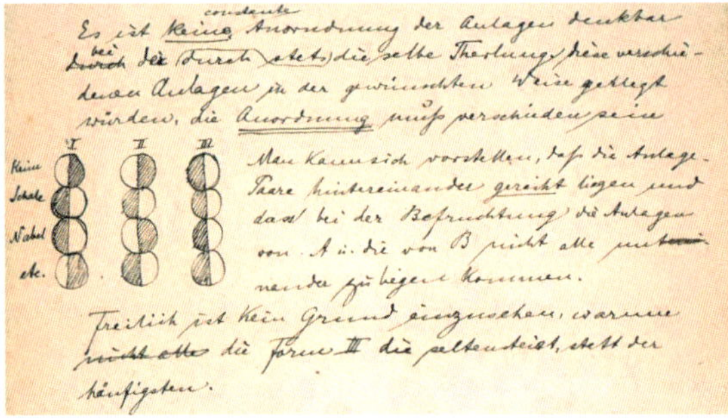

Fig. 1: Sketch from the experimental protocols of geneticist Carl Correns.

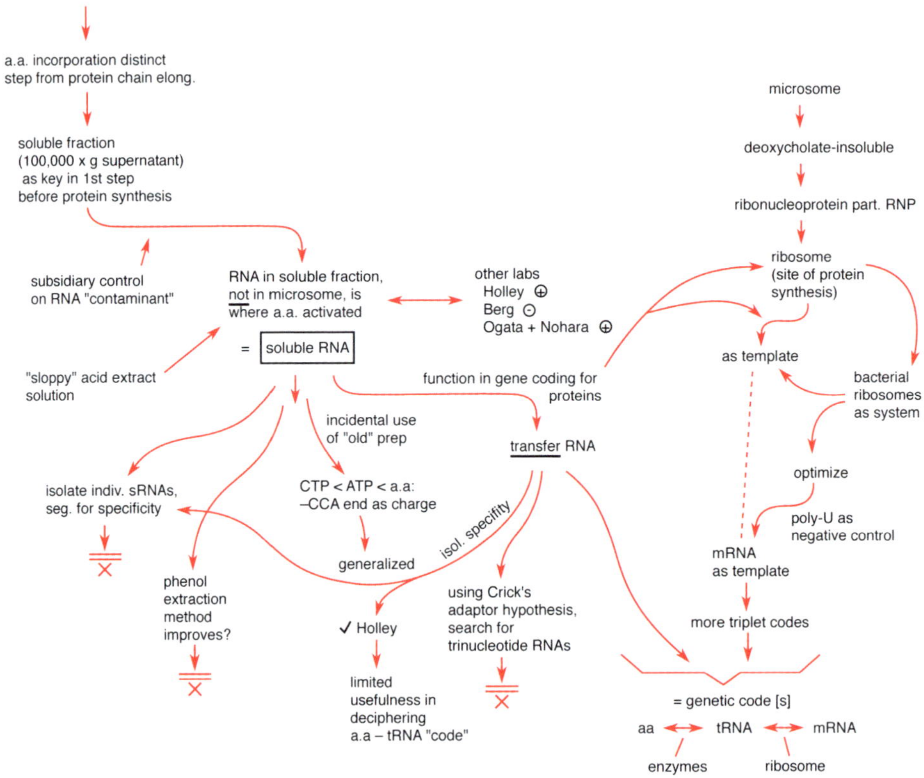

Fig. 2: Detail of protein synthesis research (1945-1960).

organism itself. The difference between inside and outside, the visible and the invisible, gave rise to the dichotomy of genotype and phenotype. The influence of this dichotomy lasted for over the whole twentieth century, and we are still living with it.

This image of the gene as the atom of life, one that was not physically graspable – it remained, for instance, unclear whether it should be addressed as a particle or a force – gave rise to endless speculation in the decades between 1900 and 1940. There was simply no way of arriving at further details by continuing with the classical cross-breeding experiments that genetics was based on in those years. It was a decisive step in this process when around the middle of the twentieth century a new generation of research technologies was developed that together allowed for breaking down the cells into their constituent parts. The new research technologies finally led to the characterization of two classes of biological macromolecules that were deemed to be crucial for the basic processes of life: proteins and nucleic acids. In the famous paper by James Watson and Francis Crick from 1953 a first image of the deoxyribonucleic acid double helix was presented. With that, a new development of the concept of the gene was taking shape in two directions over the next decade. On the one hand, genes were identified as material macromolecules; on the other hand they were seen as presenting something immaterial, bits of information deposited in the depth of the organism and responsible for the production of certain products, that is, bio-catalysts in the form of enzymes that were driving the organism.

Finally, moving to the way that today's world characterizes this realm of research, we find a third image, that of a genetic map, an order of multiple elements whose characteristics are also multiple and can be superimposed on each other. We have, then, the image of an atom of life, the image of information stored in matter, and the image of a complex network of potential processes bound together and represented in genetic maps. With current possibilities for DNA sequencing, the number of complete DNA sequences from microorganisms to mammals including humans increases every day.

To sum up, we see that in a period of about a hundred years there were several turnarounds in the conceptualization of hereditary processes. I think we can rightly say that with the concept of the gene, we have an exemplar of an epistemic concept along with an epistemic object in constant flux. Today we observe two trends in this field. On the one hand, we see developments that are leading to the reification of genes. On the day this conference paper was presented,[11] all major newspapers reported that in the United States a gene had been identified that was held to be responsible for good marriages. And recently, a gene responsible for anorexia made its appearance. On the other hand, genetic determinism increasingly gives way to an image of the organism as a much more flexible and malleable entity, as an ever-changing system of interactions, both genetic and epigenetic. Reification and liquefaction appear thus to be two aspects of the same process.

Epistemic things – the message of the example of the gene – are thus very much things in flux. Let me briefly mention another example, this time an experimental system in flux, the one I have described my book *Toward a History of Epistemic Things*.[12] Here we have a stream, a flux or a trajectory at another level (Fig. 2). The flux of the system covers roughly the time span between the end of World War II and the beginning of the 1960s. I will not go into the details here, but from the chart it can easily be seen that if we follow the path from the upper left to the lower right, we get the impression of a stream that leads us through a labyrinth – the labyrinth of experimentation. More often than not, the system generates bypaths that scientists follow for a while, until they realize that they do not lead anywhere, after which they are abandoned. We also have inputs feeding into the process in terms of new technologies, results from other experimental lines, or technical developments that are integrated into the system. Overall, the figure conveys a good image of what it means to work on and with an experimental system without being able to anticipate the direction in which the system will lead in the end. In this concrete example, Paul Zamecnik and a group of scientists at the Massachusetts General Hospital in Boston started to work on a question related to cancer research in 1945; after 15 years, they ended up with a system in which the genetic code was eventually cracked. None of the actors could have imagined that 15 years after the onset of this experimental process, something like the elucidation of the genetic code would be imminent. At its core, the path of such a research process cannot be predicted. Vagaries and indefiniteness are inherent in it. Not everything, in other words, can be anticipated. Experimental systems are a matter of tinkering; they are crafted and at the same time recalcitrant. They are proliferating, ramifying, hybridizing, and fusing together. In all of this, we see that metaphors of liquidity abound.

Liquidity as material

At this point in my presentation, I would like to switch from the more metaphorical aspects of liquidity to a more concrete mode. In the following examples, I will try to show how liquid and solid materials, immobile and mobile components, interact in a research process, and how these two modes of materiality, or phases of matter, are brought into interaction with each other. To start with, I will pick out two of the instruments that were central to the experimental trajectory just presented. The first is the oil-driven ultracentrifuge. It was developed in the 1920s and the early 1930s in Uppsala, Sweden. Theodor Svedberg was its principal creator. The original instrument is actually still there, located in the basement of one of the university buildings. The machine stands as it was abandoned in the 1950s. Obviously, the door of the lab was simply locked, the key was withdrawn, and the machine remained in the position it was in after its last run.

The interesting thing, in our context, is that this is a very heavy piece of machinery. Within this massive machine, we have a spinning rotor that is also very heavy and usually made out of the heavy metal titanium. We then have tiny holes in this rotor, and inserted into these holes are little tubes. These little tubes contain a fluid in which macromolecules are suspended. By spinning the rotor at high speeds, these macromolecules can sediment through the watery phase of the test tube. Finally, in order to use the machine for analytic purposes, one has to mount an optical device that is able to take pictures of the materials sedimenting as bands with different velocities according to their molecular weight. So there is huge and very rigid machinery here, but at its center, there is a tiny amount of liquid. Without this liquid center, the machine would make no sense at all. But you cannot even see it from outside. It is hidden deep in the machinery itself.

We have a similar situation with another technology that was very important for the experimental trajectory that I have been expanding upon.[13] It is the liquid scintillation counter devised for measuring low energy radioactive samples. Again, a very stable machine is needed for this, equipped with electronics, and able to make use of a photo-chemical process: capturing photons that are generated in the liquid material by the energy set free through radioactive decay processes. Again, the heavy piece of machinery would make no sense without a tiny amount of liquid in a vial at its center (Fig. 3). Here, as one can see, the vial is inserted into the counting-cavity by the operator. Inside the cavity, the game of liquefaction takes place, and it proves to be a very tricky procedure. In order to amplify the signal of the probe's radioactive decay process, a kind of liquid is required that is, unfortunately, only able to take up a very limited amount of water. However, as we know, the materials derived from biological experimentation are made up almost entirely

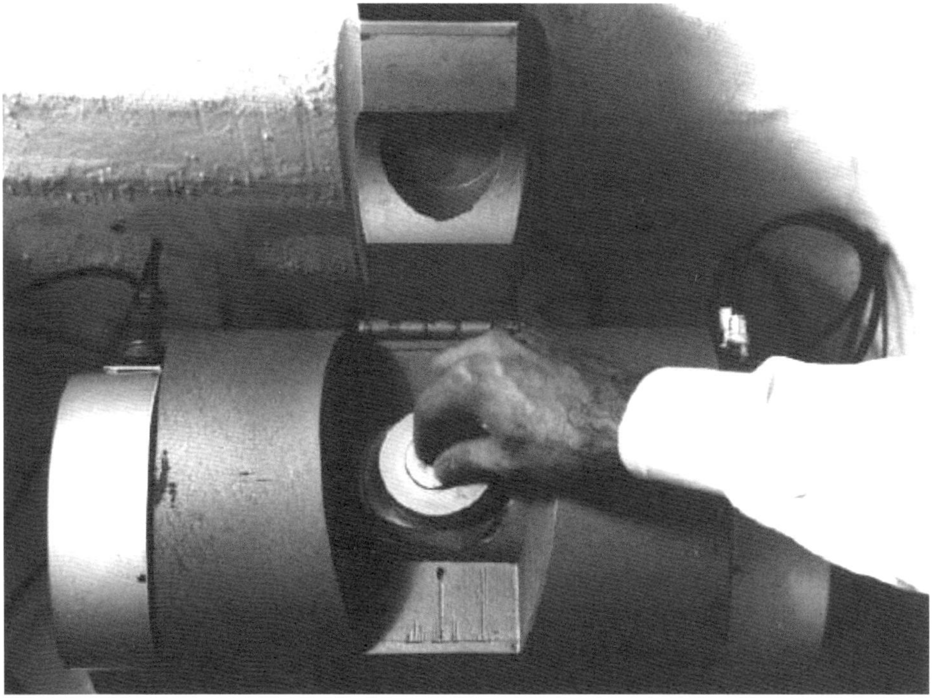

Fig. 3: Detail of liquid scintillation counter (inserting vial into counting chamber).

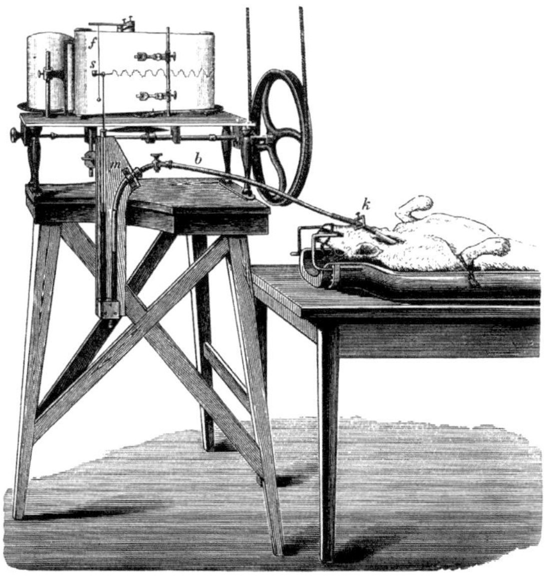

Fig. 4: Carl Ludwig's Kymograph from around 1840.

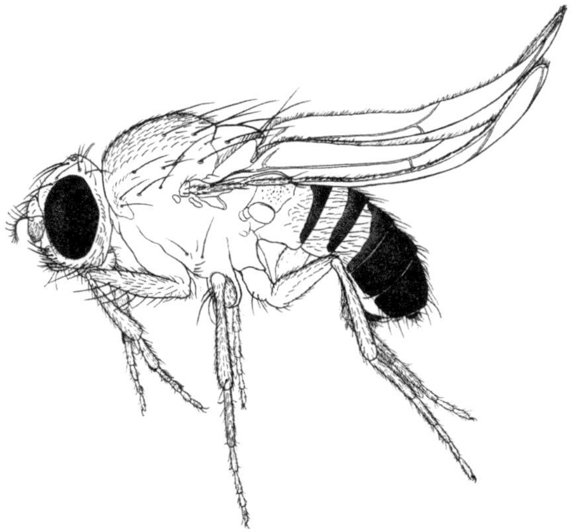

Fig. 5: Ski-winged mutant of *Drosophila melanogaster*.

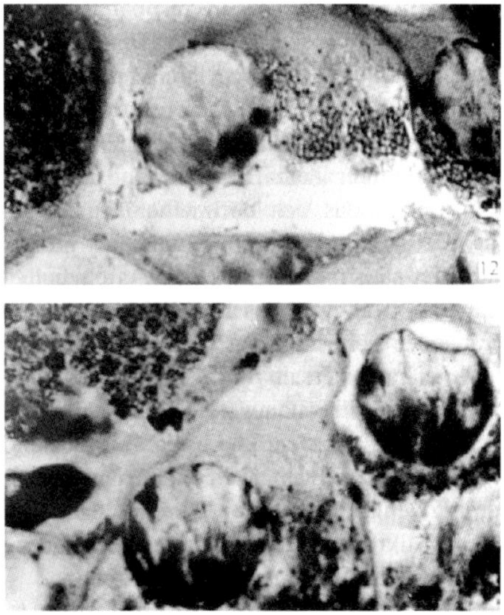

Fig. 6: Sedimentation in different layers after centrifugation of whole cells at high speed.

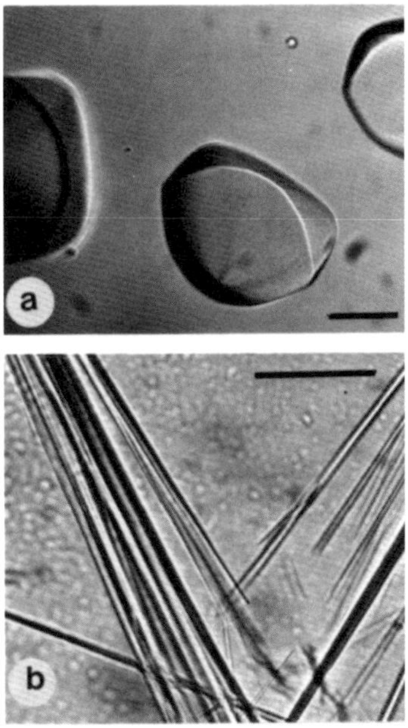

Fig. 7: Crystallized material for X-ray analysis.

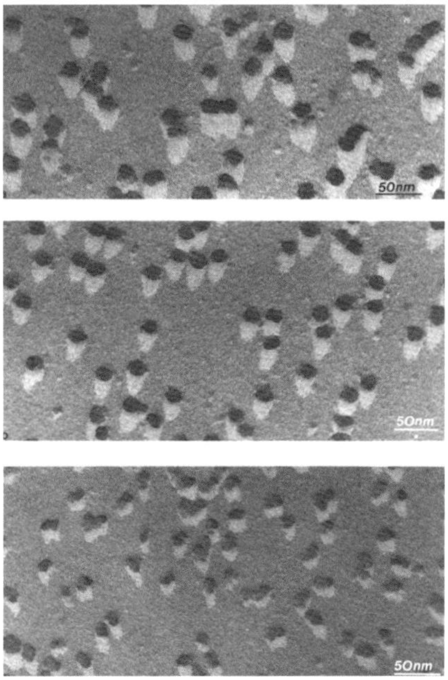

Fig. 8: Metal replica of biological material for use in electron microscopy.

of water. They need special treatment in order to mix with the solvent in the liquid scintillation counter. Again, we have a liquid and the problem of liquidity right at the center of the machinery that, if one looks at it from the outside, gives the impression of being a very heavy piece of technical and mechanical laboratory equipment.

At this point I would like to generalize these observations in a way that amounts to a negotiation of the boundary between the living and the non-living, or, to put it another way, between the soft and the hard, the wet and the dry. This boundary–depending on which experimental apparatus we look at–can assume very different and intricate shapes. I will show a few of these interfaces that are, at the same time, characteristic of certain phases of the life sciences of the past century and a half.[14] The first figure shows a kymograph (Fig. 4), an instrument that was developed in the 1840s by the physiologist Carl Ludwig in Leipzig. It can measure the blood pressure of a living animal and its temporal variations, rendered as a curve. It is a picture that would probably be forbidden in a textbook of today because it shows an animal being connected to a machinery circuit while still alive. The centerpiece is inconspicuous–a flexible tube. In this particular case, the dry and the wet, the animal and the machine, are connected with a rubber tube that itself contains another liquid, mercury, a liquid metal not miscible with water and thus with blood. The liquid metal, according to the pressure from the animal on one side, then impinges on the writing device that one sees on the other side. Finally, the blood pressure that changes with the heartbeat is recorded as a curve on a rotating cylinder. Here, the mediation between the hard and the soft occurs through a flexible link that connects the organism with a piece of research equipment.

There are very different ways to negotiate the boundary between wet and dry in the life sciences. Here is another example. What we see is a fruit fly (Fig. 5), a particular mutation of a fruit fly–its wings are so-called ski-wings. Geneticists who were working with the fruit fly *Drosophila melanogaster* in the 1920s used this mutation in their cross-breeding experiments. What we are confronted with here, and what I find interesting with respect to the use of model organisms like those dealt with in classical genetics, is that the boundaries between instrument and organism are being internalized, in a way, and acted out as the hybridization process of two different organisms. The model organisms function as soft research tools whose target is within them. We encounter such a situation not only in classical but also in molecular genetics. In the latter case, there are, for instance, petri dishes on which certain bacteria grow. If infected with different viruses, the viruses spread from an infection point and form plaques that show different shapes and, at times, different colors. Here, again, we have an internalization of the boundary between research tool and research object. We could also say that the research tool itself becomes liquefied.

On the other hand, there are research procedures in which the boundary is externalized. This is generally the case with in-vitro experimentation. Here, bits and pieces of organic matter are isolated from homogenized tissues and then experimented upon in test tubes. Instead of internalizing the boundary as in classical genetics, the boundary is externalized. The liquid is now contained in a test tube and the test tube replaces the organism. A very interesting example, in this context, is the centrifugation of whole cells at high speed, with the result that the contents of the cells are made to sediment in four different layers, as can be seen the following picture (Fig. 6).

There are, however, more and other forms of dealing with the boundary between the liquid and the solid. I will show two additional examples. We now switch to the other extreme. Here, we find instances where the boundary between the organism and the measuring environment into which it has to be placed in order to learn something about it is itself solidified. A good example for such a solidification process is X-ray analysis: to put it more precisely, the analysis of crystals through X-rays (Fig. 7). X-ray analysis needs crystallized material. A biologist who wants to find out about the structure of his or her macromolecule has to crystallize the molecule in order to apply the procedure. What is usually in a more or less fluid state within the cell has to be brought into a crystalline order to fit the research technology.

Finally, let us have a look at electron microscopy. Here, we encounter yet another kind of reification of the bits and pieces of the organism that one would like to analyze so that it can fit the machinery with which the analysis is performed. Because of its physical makeup, electron microscopy requires that the samples are exposed to a high vacuum and a strong electron beam. It is very difficult to deal with liquids in a high vacuum because everything fluid will evaporate within fractions of a second. In addition, soft materials are easily burnt by the beam. Therefore, a way has to be found to satisfy the necessity of dealing with materials from which each and every drop of water has been withdrawn – not an easy task with biological samples. As a result, it took at least as long to find a relatively stable way of handling biological samples in the electron microscope as it did to develop the entire physical machinery. Sample preparation is a crucial process in the life sciences, but what we encounter here is particularly interesting (Fig. 8).

One starts with a sample of freeze-dried biological material, but then one blasts a stream of vaporized metal over this sample so that its contours become fixed. After that, all biological leftovers are macerated away from the coat and what remains is a metal replica that is inserted into the electron microscope chamber. Of course, the aim is not to learn something about the metal that one is looking at – tungsten in the example – but to learn about the biological material that it covered, now long gone.

Thus, playing with and adjusting the boundary between the wet and the dry, the hard and the soft, the living and the non-living, can take very different forms, according to the research procedures at stake. These procedures have been subject to change over time. They continue to proliferate, and they will change in unforeseeable ways in a future that we cannot anticipate. I think, however, that we touch here at the core of the scientific exploration of the living, its limits and its prospects. As we have seen, the problems with this boundary may be circumvented, to a certain extent, if wet things, organisms themselves, are transformed into tools of their own exploration. There is fascination in the prospect that a living thing can itself act as a *phenomenotechnique* in the sense in which Gaston Bachelard has been using this term.[15] As of today, synthetic biology is a case in point. But the choices involved in this transformation will continue to be questioned as they were questioned in the past. It lies in the essence of the epistemic encounter we call research to continually measure and check the limits of its methods.

1 Kubler, George: *The Shape of Time, Remarks on the History of Things,* Yale University Press, 1962.
2 Kuhn, Thomas S.: *The Structure of Scientific Revolutions, second edition*, The University of Chicago Press, 1970.
3 Kuhn, Thomas S.: *The Trouble with the Historical Philosophy of Science*, Harvard University Press, 1992.
4 Bernard, Claude: *Philosophie: Manuscrit inédit,* edited by Jacques Chevalier, Editions Hatier-Boivin, 1954, p. 26.
5 von Helmholtz, Hermann: *Erinnerungen,* in: *Vorträge und Reden, Erster Band*, Olms-Weidmann, 2002, pp. 1-21, on p. 14.
6 Fleck, Ludwik: *Genesis and Development of a Scientific Fact*, The University of Chicago Press, 1979 [first published in: German, Basel 1935].
7 Serres, Michel (ed.): *Eléments d'histoire des sciences*, Bordas, 1989.
8 Elkana, Yehuda: *Helmholtz' 'Kraft': An illustration of concepts in flux,* in: *Historical Studies in the Physical Sciences 2*, 1970, pp. 263-298.
9 On pathways of investigation, see Holmes, Frederic L.: *Investigative Pathways. Patterns and Stages in the Careers of Experimental Sciencists*, Yale University Press, 2004.
10 For more details, see Müller-Wille, Staffan, and Rheinberger, Hans-Jörg: *The Gene: From Genetics to Postgenomics*. The University of Chicago Press, in press.
11 The date refers to October 11, 2013.
12 Rheinberger, Hans-Jörg: *Toward a History of Epistemic Things. Synthesizing Proteins in the Test Tube*, Stanford University Press, 1997.
13 For more details, see Rheinberger, Hans-Jörg: *An Epistemology of the Concrete*, Duke University Press, 2010, esp. chapter 9 and chapter 11 for the examples that follow.
14 For more details, see Rheinberger 2010, chapter 12.
15 Bachelard, Gaston: *The New Scientific Spirit* (1934), Beacon Press, 1985.

Sources of the Figures:

Fig. 1 Archive of the Max Planck Society, Section 3, Folder 17, No. 115. Courtesy of the AMPS.
Fig. 2 Trajectory of protein synthesis research at the Massachusetts General Hospital, 1945-1965. Courtesy of Douglas Allchinn.
Fig. 3 Counting chamber of an early Packard liquid scintillation counter. Courtesy of Lyle E. Packard.
Fig. 4 Carl Ludwig's kymograph. Taken from Oskar Langendorff, Physiologische Graphik. Deuticke, Leipzig and Wien 1891, p. 206.
Fig. 5 Ski-winged mutant of Drosophila melanogaster. Bibliographia Genetica. Vol. 2. Martinus Nijhoff, 'S-Gravenhage 1925, p. 42.
Fig. 6 Amphiuma liver cells spun at high speed. Albert Claude, The Harvey Lecture 43 (1950), Figs. 12 & 13.
Fig. 7 Crystals of bacterial ribosomes. The Ribosome. Structure, Function & Evolution. American Society for Microbiology, Washington D.C. 1990, p. 136. ©1990 American Society for Microbiology. Used with permission. No further reproduction or distribution is permitted without the prior written permission of American Society for Microbiology
Fig. 8 Freeze-dried and metal-shaded ribosomes. Ribosomes. Structure, Function, and Genetics. University Park Press, Baltimore 1980, p. 176. Courtesy of Georg Stöffler.

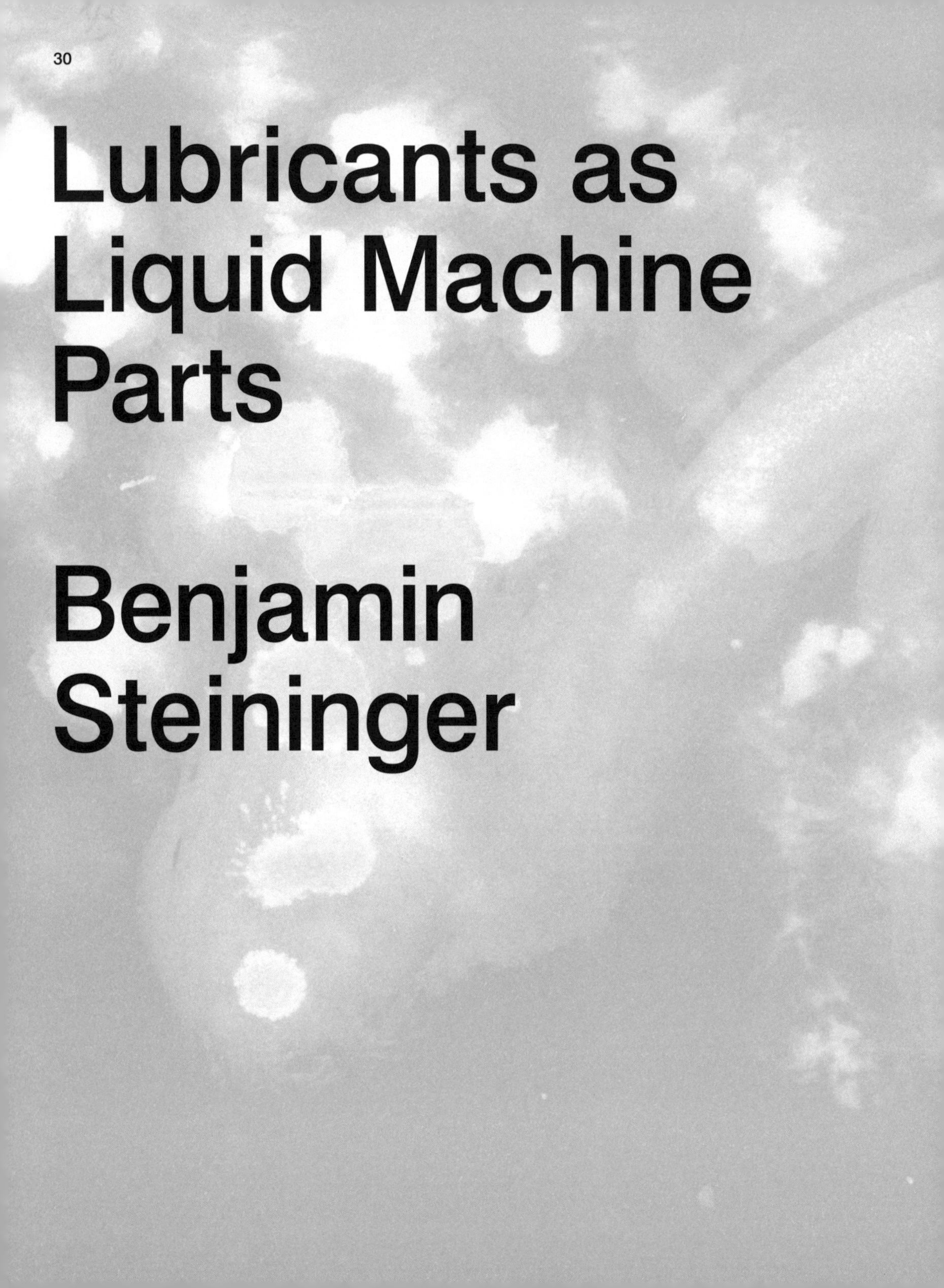

Lubricants as Liquid Machine Parts

Benjamin Steininger

"It's more than just oil. It's liquid engineering."[1] The famous slogan of *Castrol Motor Oil* presents motor oil as a product of high technology. But as the name of the company reveals, this technology of liquid lubricants is based on materials that typically do not play a role in engineering. In fact, the brand name *Castrol* is derived not from a high-tech material at all, but rather from a natural substance: *castor oil* (from the Latin *oleum ricini sive castoris*). A versatile substance, castor oil is relied upon by a variety of industries, occupations, and sciences to serve multiple, and often divergent, purposes. For example, on the one hand, it acts as an excellent laxative for inducing childbirth; while on the other, it is a prominent source of ricin, a highly toxic chemical. However, one of its most easily recognized applications is as a powerful and historically unrivalled engine oil, especially for high-speed race car engines.

Despite their mostly single-minded profitmaking approach, engine oil companies and related brands illustrate, via their advertisements, that engine lubrication is a topic around which very different material cultures intersect. This essay presents an approach to the influential–and often strange–materiality and agency of lubricants, from a perspective informed by cultural history and media theory. At least since Sigfried Giedon's (1888-1968) famous observation that "the sun is mirrored even in a coffee spoon",[2] a scholarly tradition has existed that focuses on the small, neglected parts of the technological system. Lubricants are a prime example of such neglected parts, which is unfortunate given their colorful history. As liquids, they are obviously materialistic outsiders in the world of machines. But, as such, they create special bonds between chemistry and mechanical engineering, between solid, especially metallic and liquid materials, and between fossils of long-dead creatures and their living counterparts.

1. Materials in Cultural History

The *Stone* Age. The *Bronze* Age. The *Iron* Age; each of these epochs and their cultures are defined by the dominant materials they used. From the forge to quantum mechanics, materials of every type have impacted the course of human history and its institutions; and, conversely, these institutions (technological, cultural, and scientific, to name but a few) have been historically invested in the development and application of material resources. In modern times, however, the rise of chemical industries has complicated this picture. Crude oil, rare earth elements, and other materials are both paradigmatic technological resources and matters of global concern and conflict. To single out one material as paradigmatic of the twentieth century would be impossible. Do we live in a hydrocarbon age or a nuclear age? Is ours the age of biochemistry, nanotechnology, or even information–and if so, would this mean we are actually living in a *post*-material age? Plastic or concrete, silicon or steel, DNA or CO_2, oil, uranium or even coal–any of these could serve as an accurate starting point for talking about modern materiality. And yet, discussing any one of them would inevitably lead us into a discussion of the rest. Similar questions about nature and industry, about global and local phenomena, about molecular and planetary knowledge, about the economic, social, scientific and political agency of materials, both new and old, could emerge almost everywhere. Thus, the theoretical framework required for an interpretation of twentieth century technology would need to incorporate not just one material or industry but an ensemble, both old and new, and in all possible states (i.e. solid, liquid, or gas). And it is not just the most prominent materials that should be considered as meaningful representations of twentieth century material culture. On the contrary, *pars pro toto*, the small, neglected, and obscure parts of global material culture can also tell us stories that will help resolve theoretical problems, if we are but willing to consider them.

2. From the Materiality of Media to the Mediality of Materials

The concept of *materiality* has attracted significant academic attention over the last few decades, especially for scholars of the history of science, media studies, literary studies, and art history. In fact, so intense has their scrutiny been that a new term had to be coined to describe it: *the material turn*. In a wide range of disciplines (most of which are historically oriented or humanistic), it has become obvious that material, practical foundations underlie the process of communication and the otherwise abstract history of ideas and concepts.[3] The development of media studies as a discipline was itself predicated on such a realization. Research on the material and historical basis of literature and philosophy has shifted to research on technical media. Despite this trend, however, there has still been comparatively little academic interest in the study of the materials themselves.

Instead, the lion's share of research associated with the so-called material turn has focused mainly on the material apparatus of the sciences–from printing to computing, from the kitchen to the laboratory–and the sociohistorical practices related to them. To be fair, some cultural histories of materials have been produced: *biographies* of water, aluminum, wood, and iron, to name a few. But toward a more comprehensive theory of media, there remains much theoretical and research potential. Of all the humanities, media theory obviously has the most technical interest in culture. Since the very beginning of academic media studies, researchers such as Harrold Innis (1894-1952)[4] and Marshall McLuhan (1911-1980)[5] have *smuggled in* topics directly related to material structures (e.g., roads, electrical grids, transportation methods) into discourses about art and writing. In German academia, and especially under Friedrich Kittler (1943-2011) and the Berlin School, the material principles of a wide range of very different machines and apparatuses, from air pumps to microscopes, from the *cithara*[6] to the *Enigma machine*[7] to the microprocessor, have driven the interests and agenda of the discipline.[8] This emphasis on *technical* materialism, and the epistemological principles on which it is founded, has led to a concept of media that goes far beyond conventional theories of mass media, whose propositions were designed with respect to the social impact of e.g. television, news, and the Internet. According to a technical materiality perspective, both technology and science as a whole fall under the research purview of media studies. If our aim is thus to show how technical systems shape and define abstract theoretical concepts, then there can be no strict boundaries between this version of media studies and epistemology. A technologically enhanced version of media studies can contribute to the work of epistemology; and conversely, scientific systems, be they historical or contemporaneous, can also contribute to a reflection on the media themselves. And if we aim to analyse the mediality of materials, a second enhancement of the concept of media seems fruitful. Not only the processing of information, but as well the processing of energy should be analysed.

Physical materials are manipulated and applied by science and technology in a multitude of ways, and understanding those ways can serve to shape and sharpen typical media concepts like transmission, storage, and processing. In fact, some of the most productive principles of media theory are embodied in and expressed by materials. Sometimes, more than one of these principles are evoked by a single material. Fuels, for example, are used to *transmit* energy from one place to another; but they also represent, by their very nature as fuels, a form of chemical energy storage. Crude material manifestations of abstract concepts like transmission or storage can enhance this understanding because of the richness of the material sphere. There is always a certain surplus in material reality and thus also the conceptual framework can be sharpened.

Chemistry and the history of materials have not yet been researched in enough detail to warrant academic debate in media studies. However, from well-established research on the materiality of media, the mediality of materials can be addressed.

Some materials literally work as media: they transmit signals and energy. This is as true for biology as it is for technology. Take biochemistry: hormones, for example, transmit *signals* and *messages* to cells; biocatalysts form *intermediate* compounds to accelerate chemical reactions; enzymes, a type of biocatalyst, operate on the *key-lock principle*, fitting together with other molecular structures to trigger specific reactions. Certain materials such as water or other solvents are considered as dispersion media in chemistry, other materials like agar are used as nutrient media, as a certain milieu, in which bacteria can grow, in biology. The specific mediality of materials such as these is comparatively little discussed in media theory. And yet the mere question – in what sense do materials serve as transmitters in natural or technological environments, or as storage or processing units – could both enrich and enlarge our conceptual understanding of media as a whole. This essay attempts to do just that, in the context of one particularly important material: engine lubricants.

3. Kinetic Transmitters at the Heart of the Machine

Lubricants are intermediate materials. Kinetically, they work as transmitters of energy. Already as such, lubricants appear as something like a material medium. But where does this perspective lead to? What could we learn in general about materials in technology here? Along with a systematic and historical discussion of the materiality of lubricants, I will also contribute to the literature on the scientific and cultural agency of materials in general, and of liquid materials in particular.

The materiality of most common machines is rarely questioned. Nothing could be more obvious: machines are made of metal. While they rely on a solid (e.g. coal), liquid (e.g. oil), or gaseous (e.g. natural gas) fuel to function, they nonetheless consist of solid parts, the arrangement and design of which also enables them to function. However, even simple machines like bicycles will inevitably break or otherwise fail if not *lubricated* properly. At first, bicycle chains would creak and grind; but sooner or later, friction between the dry moving parts would degrade and weaken them, until eventually the whole assemblage would break down or apart. This process is even more severe with car engines. Even the most mechanically illiterate people know full well not to drive their vehicle without engine oil. Simply put, lubricants are *essential* for solid machinery to function. But what does this conventional knowledge imply? How does lubrication actually work? What substances are they made from? And, for the purposes of this essay, what does our reliance on these substances tell us about modern materiality in general? What types of knowledge and skills are involved with the production and application of lubricants?

As the Castrol slogan state–*it's liquid engineering*–lubricants like engine oil are mechanical engineering–they are, in essence, *liquid technology*. It is upon this machine-oriented framework, and the theoretical principles it evokes, where our discussion of lubricants must necessarily begin. Within academic theories of media, the phenomenon of machines is comparatively well discussed, ranging from concrete kinetic machines to abstract paper machines to program-controlled electronic devices. One of the most influential typologies used to understand machine systems is the contrast between *closed* and *open* systems for processing kinetic energy or information. The degree of closure of a system can almost be considered a criterion for its technological character. Biological organisms, on the other hand, are capable of adaptation–in other words, *openness*–to environmental changes. Machines, however, are typically built for a specific purpose or environment, and cannot be deployed for new purposes or in different settings without risking damage or destruction.

The concept of closed and open machine systems was first used by Franz Reuleaux (1829-1905), a professor of mechanical engineering at the Technical University of Berlin (in his time, it was known as *Technische Hochschule Charlottenburg*). According to technology historian Peter Berz, Reuleaux, in his conceptualization of the kinematics of machinery and his corresponding axiom that "Machines, that work, are closed systems" essentially "deduced in the last quarter of the nineteenth century nothing less than a science of mechanical functions."[9]

With the emergence of cybernetics and the related concept of feedback, open machine systems became possible. But in the strictest sense, technological systems that interact autonomously and productively with their environment (i.e. open systems) have not existed until recently. Biological organisms, on the other hand, are open systems by nature. Whereas machines are built to ignore their environment as much as possible, biological organisms both emerged from and are sustained through continuous interaction with their environments. As such, the contrast between open and closed systems is a useful dichotomy to consider for the analysis of technology.

For the purposes of our discussion of lubricants, we must transfer this concept. We must transfer the analytical contrast between closed and open systems, from the economies of machines – be they kinetic or energetic – to their materiality. Also the set of materials used in mechanical engineering can be interpreted as either an open or closed system. Obviously, lubricants indicate open ends of a kinetic system. But as liquid things they could also indicate open ends of the material system of metallic machinery.

I began this discussion by defining my conceptual interests in materiality, and then reviewed some of their tangible manifestations. And while my approach to materiality is essentially systematic, it must also be situated historically: as a progression from timeless principles of materials to their concrete historical expressions. Only by considering both avenues of thought simultaneously will the conceptualization of materiality be complete. To that end, an historic finding from 1938 will serve as the starting point for this part of the discussion.

4. Walter Ostwald's Lubrication in Transition (Schmierung im Wandel)

In 1938, a short article titled *Lubrication in Transition (Schmierung im Wandel)* appeared in *Automobiltechnische Zeitschrift, (Automotive Engineering Magazine)*.[10] It was written by Walter Ostwald (1886-1957), a German fuel chemist, automobile journalist, patent holder, and inventor of *Aral*, one of the most popular car fuels in Germany during the twentieth century. His father, chemist Wilhelm Ostwald (1853-1932), was prestigious in his own right, earning a Nobel Prize in 1909 for his historic work on catalysis.

Importantly, Walter Ostwald observed that "in the field of lubrication almost everything is in flux."[11] He also enumerated the significant technical details involved in researching lubricants:

> *Things like inlet lubrication, colloidal graphite, h.p.-lubrication pins, viscosity index, active molecules, resistance to oxidation, freedom of corrosion, formation of emulsions, additives to lower the point of solidifying, "oiliness"[12] […] and many other questions of lubrication are currently burning. And every single one is the subject of controversial statements and experience.*[13]

Open questions burned everywhere, but almost nowhere was solid knowledge obtainable:

> *Even the theory of lubrication and the related experimental research provide new facts and aspects that strengthen the perception that we currently learn a lot of new things but that we know depressingly little for sure.*[14]

Ostwald continues:

> *This state of a science field can be quite unpleasant for someone who works in manufacturing, who can't wait for perfection of our knowledge. But such a state – one can add: on a massive scale, open questions in technical fields, lack of theoretical instruments – is always announcing forthcoming progress.*[15]

But in fact, this state also provides the opportunity for fundamental reflections and statements about the *essence of lubrication*:

> *Lubrication is a strange natural process, because the lubricant works only by its presence, without being consumed. The process is closely related with catalysis in chemistry, where the mere presence of certain substances causes chemical reactions that otherwise would occur with different speeds or different directions. Chemical reactions are based on intermediate compounds of the catalyst, and similarly in the case of lubrication, direct friction is replaced by intermediate friction, from solid to solid to liquid to liquid (in the limiting case). Similar to catalysts that increase the yields and repress side*

> *reactions, also lubricants increase the energetic effectivity of an engine, and reduce the amount of work that is withdrawn in the form of heat loss.*[16]

Observations like these, in a technical journal, provoke several questions. Why this parallelization of lubricants and catalysts? What does this tell us about technological discourse? Is this merely wordplay?

When put in context, we can see that Walter Ostwald deliberately refers this catalytic rhetoric to another one. He is, in fact, just reversing another metaphor. In his father's influential 1893 work, *Lehrbuch der allgemeinen Chemie (Textbook of General Chemistry)*, we find several sections that discuss catalytic processes. In the chapter Electromotive Forces, for example, Wilhelm Ostwald writes:

> *Solutions of salts, of copper, antimony, silver, or similar metals cannot be reduced by gaseous oxygen … But if one adds a platinized sheet of platinum to the solution in contact with hydrogen, precipitation of the dissolved metal takes place at the platinized sheet, and hydrogen takes its place while the corresponding acid is formed.*[17]

In the following section, he writes, "We attribute this effect – without a real explanation – to the 'catalytic force' of the platinum." Wilhelm Ostwald may not provide a *real explanation*, but he does offer a *telling comparison*:

> *The platinum works like ice on the side of a hill, or the lubricant in a machine. Without them, even given the same motive forces, neither the sleigh on the hill nor the wheel on the dry shaft is put into motion. But if ice or a lubricant is present, the same forces cause the motion and the device will seek a state of lower energetic intensity.*[18]

From here, we find the metaphor of a catalyst being *something like* a chemical lubricant being used in several textbooks and encyclopedias published around the beginning of twentieth century.

The question is, what happened between the publication of both texts, the *Textbook of General Chemistry* in 1893 and *Lubrication in Transition* in 1938? In both cases, one technical field was explained in terms of another, and the obscure explained in terms of the more obvious. How machines are lubricated seemed self-evident in 1893, and how platinum works, still strange. The model that helps Wilhelm Ostwald's contemporaries understand chemical dynamics is taken from the contemporaneously successful sphere of industry, and from the kinetics of machinery.

In 1938, the direction of the explanation had changed. Among all of the working parts of machines, among all of the *technical objects*–to use Hans-Jörg Rheinberger's expression–new areas of unknown phenomena, new *epistemic things*, appeared.[19] The formerly self-evident world of machines now required a new explanatory metaphor, this time taken from the field of chemistry. Not only resemble the microspheres of catalysis and lubricants in a metaphoric way. For Walter Ostwald at least, the technology of catalysis provides also a model for depicting industrial scientific success. In just a few decades, between 1890 and 1940, catalytic technology had evolved from a mysterious yet marginal scientific topic to one of the most powerful foci in the technosciences. Historically, this technology came of age after mechanics but before the Atomic Age, after steam engines but before cybernetics. This is important to identify, as it underscores how the chemical, catalytic industry acted like a bridge between the industrial paradigms of the nineteenth century and the technologies and related lifeways integral to the second half of the twentieth century.[20]

Although explaining catalysts as lubricants and lubricants as catalysts might seem to be a play on words, it could also be interpreted as a symptom of a dialectical play in the evolution of industrial processes and technologies, from mechanical engineering to chemistry–and from there, to hybrid fields in-between.

From a historical perspective, the use of unconventional substances to produce machine lubricants–the very definition of *liquid engineering*–is still a nascent branch of technology. During the nineteenth century, mechanization could proceed without precise scientific knowledge about the nature, composition, or performative capabilities of lubricants produced from animal grease or vegetable oils. Steam engines or water-driven systems, for instance, were lubricated by laymen, not academically trained chemists. And by today's standards, the machines of that era were relatively straightforward and simple in design. If the machine would become hot or noisy, simple steps could be taken to fix the problem, animal grease or vegetable oil could be used to reduce friction. With the emergence of the combustion engine, however, fossil substances came into use as well. Products made from crude oil proved to be an effective means of modulating the higher temperatures, pressures, and friction generated by these new and more powerful engines. These effects became an academic subject of physical theory in the last decades of the nineteenth century, due to the work of physicists like Nikolai P. Petrov (1836-1920), Arnold Sommerfeld (1868-1951), and Osborn Reynolds (1842-1912). But as late as the 1930s, fuel chemists were still writing about engine lubrication as a widely open research area. The chemical and material aspects of lubrication seemed especially underinvestigated. In 1936, the first American book on the subject, *Theory of Lubrication*, was published by Mayo Hershey (1886-1978).[21] The reaction his book received could best be described as uninformed if not woefully ignorant: "Why a whole book on lubrication? Just put in enough oil!" was one noteworthy response.[22] Within two years of its publication, however, the book had to be reprinted, demonstrating that the need for a more comprehensive understanding of lubricants was indeed warranted.

Three factors were essential to the historic, material development of the field: (1) new machines and engines, (2) new raw materials, and (3) new means to process and refine them. The speed of the steam engine was the first problem to be addressed, followed by the intense heat and friction generated by the combustion engine; this lead to the search for new lubricants, fossil substances in particular, that could modulate the effects. In parallel chemical and industrial tools were developed to synthesize and shape the lubricants up to a molecular level in order to amplify their potency. As we learn from Walter Ostwald, around 1940 new challenges emerged with respect to engines that could only be resolved with advances in chemical engineering.

5. From Function to Constitution: The Materiality of Lubricants

In *Lubrication in Transition*, Walter Ostwald described lubricants as a means to *seal machines energetically*.[23] Simply put, lubricants make machines lose less energy. So, even if they do not function as fuels themselves, they nonetheless remain crucial for the energetic effectiveness of a mechanical system. Thus, friction-reducing materials do provide an engine with energy, regardless of whether or not they serve as the actual fuel source. As Anton Zischka (1904-1997), an eccentric but–for a certain technopolitical discussion–symptomatic Austrian journalist and author of many books on the historic impact of energy and raw materials, once stated: "The giant powers of Niagara Falls are small compared to the powers that are saved by the silvery steel balls that facilitate the course of almost every machine."[24]

To close a machine system, kinetically or energetically, means to open it in a material sense. To seal machines energetically takes far more than just Zischka's *silvery steel balls*. The metallic gears, threads, screws, ball bearings, piston rings, cylinder surfaces, and other parts comprising a machine are in constant contact with each other, and as such, lubricating materials (e.g. fats, oils, greases) must inevitably be used to lessen friction and promote performance. To this end, a wide and often bizarre array of materials and ingredients have been suggested over the years. Substances as neatsfoot oil (*Klauenöl*), tallow (*Unschlitt*), or colza oil (*Rüböl*), are suggested in Hugo Horwitz's (1882-1941) 1914 dissertation *Evolution of Bearings (Entwicklungsgeschichte der Traglager)*[25] or in *Auto Chemistry (Autler-Chemie)*,[26] Walter Ostwald's 1910 do-it-yourself book on car repair .

Even in 1940 – and this means during a widely motorized World War –, a range of unconventional lubricants were still being seriously considered, even in highly regarded German technical journals on high-speed aviation.[27] Engineers and scientists conducted experiments on a diverse range of materials, including olive oil, castor oil, and sperm-oil (made from sperm whales), to name a few. However, it is also clear that at this time, more chemically informed and experience-based research on lubricants had begun to fully develop.

6. Oiliness – Closed Systems are Open Systems

One term that could lead to a more detailed interpretation of the opening of the materiality of machine systems is *oiliness*. Despite being an English term, oiliness is also used in German publications, where it is translated as *slipperyness* or *lubricity*.[28] At first glance, the concept seems like an absurd self-referential term. Ostwald's terminological narrowing of lubrication and catalysis had already tried to explain one thing in terms of another, and vice versa. But here, the viscous vicious circle is set in only one word: The perfect oil is defined by its oiliness!

In both a material and historical sense, the concept of oiliness describes the opposite of a closed system. It was first deployed around 1900, to serve as an opposite yet important complement to the term *viscosity*. Viscosity was the central concept and the single physical constant of the hydrodynamic theory of friction. But in practice, viscosity could not sufficiently describe the performance of a lubricant. It often happened that certain oils had the same measurable viscosity, and yet radically differed in terms of efficiency. It became increasingly obvious to both practitioners and experimenters that physical appearances alone were not enough to accurately interpret these materials and their agency. The materiality itself–the chemical behavior of lubricants–had to be explored.

It was at this point that a pivotal distinction was made: The best lubricity was not provided by chemically *stable* substances. Fossil paraffins, ancient substances that no longer chemically react with the environment, are less efficient lubricants than mixtures made primarily from animal or plant matter, which are both chemically unstable and corruptible material.

And this chemical corruption affects more than just the lubricants themselves. As Walter Ostwald noted:

> *As general experience shows, the most effective lubricants, like vegetable oils or animalistic fats, chemically attack metal. And we know as well, that purely paraffinic–that means chemically inert–hydrocarbons are in no way good lubricants [...] With a small addition of organic acids one can heavily increase the lubricity of mineral oils. There is no doubt that a good lubricant needs "active" molecules, with aggressive "heads", that bite into the metal layer and form something like a fine fur, made of rectified microscopic molecular hairs, that is the base of the remaining process of lubrication, and that can almost not be pushed through by an imminent contact of metal to metal.*[29]

The engineering of lubricants must walk a thin line between a material's durability–its chemical age-resistance–and its required corruptibility by metallic surfaces. To reach a compromise, mixtures of different materials are tested. Paraffins are no more than base materials that must be blended with other substances. In the German language, the contemporary term for such mixtures is remarkable: *Legierungen*, or *alloys*. This term should clearly be familiar to metallurgists, blacksmiths, or anyone else who works with metals.[30]

But to create an alloy at the level of a lubricant does not mean creating a more stable liquid, but rather its opposite. Lubricants represent the open parts of machines; they occupy a position in which kinetic chains are fragile and the loss of

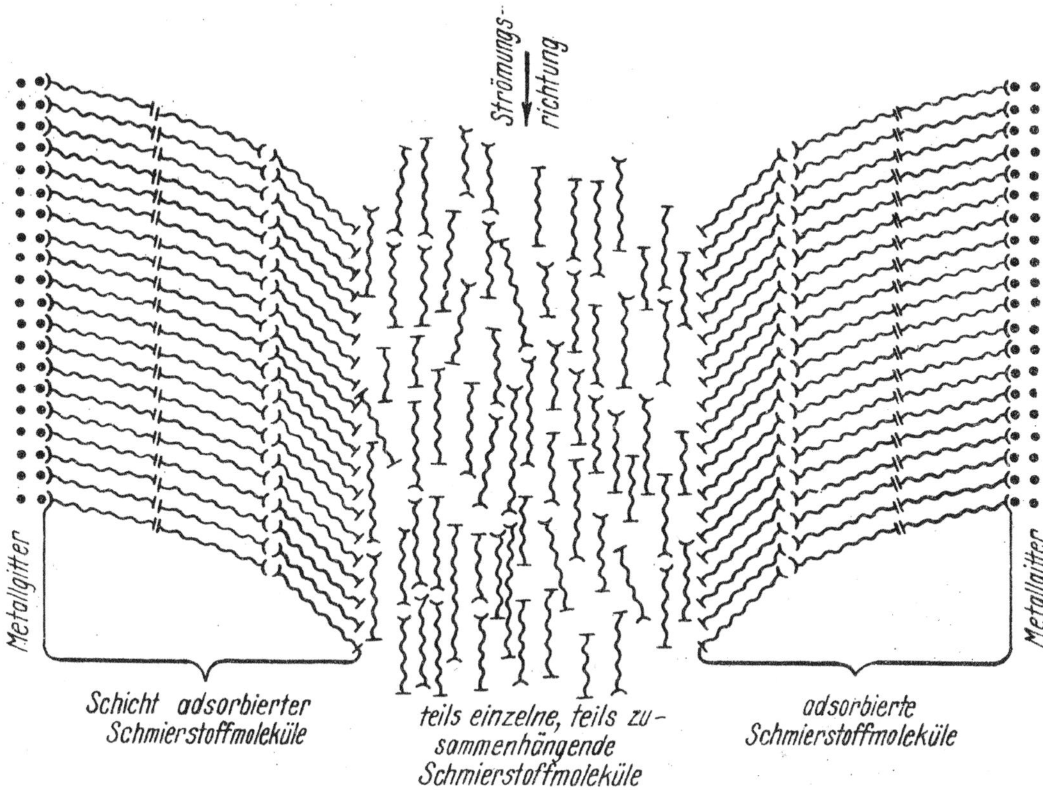

Fig. 1: Hybrid matter inside the machine: Lubricant molecules and metal surface molecules form a tiny interlayer

energy is always imminent. To function properly in an open position, the lubricant itself must be opened in a chemical sense. The substance must be made more reactive by the addition of an *Oilinessverbesserer*: a chemical substance that destabilizes both the lubricant and the solid surfaces it is in contact with.

In terms of the theory of machines, what all of this means is that machines work as closed kinetic systems in the mesosphere because they are open chemical systems in the microsphere. The parts of a well working machine no longer have clear boundaries between their metallic and their liquid parts. Some thin layers on the surface of the metallic parts of a machine belong to an intermediary space, or an intermediary materiality. They become both a part of the metallic, solid structure and a part of the liquid materiality constituting the lubricant. In a molecular sense, these surface/lubricant structures are a compound of both material spheres. The mechanical process, and even more so the chemical process, has changed both components by their interaction. Some elements of the metallic surface can even be destroyed by the lubricant, which could in turn cause both mechanical and chemical problems. Because of the catalytic activity of the metal parts in contact with the lubricant, its fragile stability as an explicitly unstable substance can suffer.

What could be understood as a certain mediality of materials is obvious here in an exemplary sense. Lubricants literally act as transmitters because they are hybrids and material media themselves.

7. Industrial Closure

On the microscale, machines function as closed systems because they are open at the level of their liquid lubricants. But on an historic scale, the heuristic impact of olive oil, neatsfoot oil, and others as a means of opening fossil lubricants in a chemical way did not last too long. Mechanical or chemical parameters such as thermal stability and oiliness are from the 1930ies not just governed by real natural substances, but increasingly by, as Walter Ostwald put it, *supernatural*,[31] and by this is meant *synthetic*, materials. And here we return to the play on words, to the parallelization of lubricants and catalysis mentioned at the beginning of this historical narrative.

The chemical industry started around 1940 with catalytic means to synthesize almost every natural substance, that can be used as a lubricant or to blend lubricants.

Thus, the experiments with olive oil or sperm-oil described in journals from that period demonstrate the emergence of diagnostic tools, while the answers to more fundamental and functional problems can be found in chemical technology and true liquid engineering. Knowledge from layworkers, who had firsthand experience blending and mixing different materials, while so impressively presented in the writings of Walter Ostwald and even in the *Ringbuch der Luftfahrttechnik*, has been replaced by high technology, academia, and industrial laboratories. As we can discern in a retrospective talk given by I.G. Farben chemist Hermann Zorn (1886-1983) in 1958, the first industrial experiments aimed at producing purely synthetic lubricants were carried out as early as 1930.[32] During the Second World War, procedures and products were urgently perfected. Synthetic lubricants that remained liquid even at temperatures around -50°C where produced for the frontlines enduring the harsh Russian winter in 1942. With industrial chemical means, frost was overcome in aircraft engines, rail wagons, and machine guns.[33]

So if lubricants still have to be considered outsiders in the workings of the machine, then since 1940 as products of another industry. With this shift their technological status has changed: they do not represent nature at her most irritating and perishable materiality, like animalistic grease or vegetable oil. They are fully technological, scientifically controlled products, whose purpose allows machines to function, even if they make them leak and get dirty.

1 Castrol's official slogan, on their website: URL: http://www.castrol.com/ accessed June 9 2014.
2 Giedion, Sigfried: *Mechanization Takes Command*. Norton 1969, p. 3
3 see, Gumbrecht, Hans Ulrich, Pfeiffer, K. Ludwig, (eds.): *Materialities of communication*, Stanford University Press 1994
4 Innis, Harold: *Empire and Communications*, Clarendon Press 1950
5 McLuhan, Marshall: *Understanding Media: The Extensions of Man*, McGraw-Hill 1964
6 Greek string instrument, ethymologically related to guitar and sithar as well. Epistemologically, citharas and the harmony of strings are related to pythagorean number theory and thus to mathemetics as such.
7 German encryption device during World War II. It was the decryption of its mechanism by British intelligence mathematicians that gave rise to modern computer history.
8 See, for example, Kittler, Friedrich: *Grammophone, Film, Typewriter*, (transl. by Geoffrey Winthrop-Young & Michael Wutz), Stanford University Press 1999; Siegert, Bernhard: *Relays: Literature as an Epoch of the Postal System*, (transl. by Kevin Repp) Stanford University Press 1999; Berz, Peter: *Ein Standard des 20. Jahrhunderts*, Fink 2001; Kittler, Friedrich & Gumbrecht, Hans Ulrich: *The Truth of the Technological World: Essays on the Genealogy of Presence*, Stanford University Press 2014
9 Berz 2001, p. 10
10 Ostwald, Walter: *Schmierung im Wandel*, in: *Automobiltechnische Zeitschrift*, 41 Heft 14, 1938, p. 365-368
11 Ostwald 1938, p. 365
12 English also in the German text!
13 ibid.
14 ibid.
15 ibid.
16 ibid.
17 Ostwald, Wilhelm: *Lehrbuch der allgemeinen Chemie (Chemische Energie)*, Engelmann 1893, pp. 900ff
18 ibid.
19 See Rheinberger, Hans-Jörg: *Toward a History of Epistemic Things: Synthesizing Proteins in the Test Tube*, Stanford University Press 1997. For a condensed version, see the encyclopedia entry at http://vlp.mpiwg berlin.mpg.de/essays/data/enc19, accessed November 4 2015. Also see Rheinberger, Hans-Jörg: *An Epistemology of the Concrete: Twentieth-Century Histories of Life*, Duke University Press, NC 2010

20 For a general account of the industrial history of catalysis, see Steininger, Benjamin: *Refinery and Catalysis*, in: Rosol, Christoph, Klingan, Kathrin, Scherer, Bernd et al. (eds.): *Textures of the Anthropocene, Grain/Vapor/Ray*, Vol. Vapor, Revolver Publishing 2014, pp. 105-118
21 Hersey, Mayo D.: Theory of Lubrication, Wiley 1936
22 Hersey, Mayo D.: *Theory and Research in Lubrication. Foundations for Future Developments*, Wiley 1966, p. 6
23 Ostwald, Walter: *Kraftstoff!*, in: *Kraftstoff* 15, 1939, p. 4.
24 Zischka, Anton: *Sieg der Arbeit, Geschichte des fünftausendjährigen Kampfes gegen Unwissenheit und Sklaverei*, Leipzig 1941, p. 299
25 Horwitz, Hugo Theodor: *Entwicklungsgeschichte der Traglager*, Buchdr. Gutenberg 1914
26 Ostwald, Walter: *Autler-Chemie,* R.C. Schmidt & Company 1910; The almost unknown german word Autler was a contemporary way to describe the motorist analogously to the cyclist (*Radler*).
27 Reichsluftfahrtministerium (ed.): *Ringbuch der Luftfahrttechnik*, Berlin-Adlershof 1940
28 See Kadmer, E. H.: *Laboratoriumsprüfung der Schmierstoffe*, in: Reichsluftfahrtministerium 1940, Vol. 4, C 2, p. 11
29 Ostwald 1938, p. 366
30 Quote after: Zima, Stefan: *Kurbeltriebe*, Vieweg 1998, p. 316; originally: Richter, W.: *Die Schmierung von Diesel und Ottomotoren unter Berücksichtigung verschiedener Brennstoffe,* in: *Brennstoff und Wärmewirtschaft,* 20, 1938, p. 4
31 Ostwald 1939, p. 4.
32 Zorn, Hermann: *Chemischer Aufbau und physikalische Eigenschaften der Schmierstoffe,* in *Bericht über die Schmiertechnische Tagung*, Wien 1958, p. 17
33 ibid., p. 21

Source of the Figure:

Fig. 1: Illustration from: Zorn, H.: *Die chemische Technologie der Schmieröle*, in: Reichsluftfahrtministerium 1940, Vol. 4, C 9, p. 30

Surfing the Waves: The Roles of Marker Materials in Turbulence Experiments

Inge Hinterwaldner

Introduction

This contribution aims to give some idea of how marker materials, which have been used in fluid dynamic experiments mainly since the nineteenth century, are processed to further the envisioned goals. What roles can be identified from the way that they are applied? By chance, air and water, the two fluids of greatest interest,[1] are both transparent. In order to gain an initial insight into the flow patterns, scientists have to make visible the events occurring in the studied fluid. For this purpose, scholars have tested a variety of materials for use as markers. With materials used as markers, in most cases we are not only dealing with existing substances that are visible and have approximately the same density and viscosity as the liquid under study. But they also often fulfill other criteria and can serve multiple purposes that include the following: being, embodying, representing and preparing an object of research, stabilizing an interface by carrying another marker, helping define visibility sections, working with both light and dark (being photogenic), and being the base for further visualizations. Although this list is far from exhaustive, it should suffice to make an argument for what the political theorist Jane Bennett calls *thing-power*. By this, she means "the capacity of things [...] not only to impede or block the will and designs of humans but also to act as *quasi agents* or forces with trajectories, propensities, or tendencies of their own."[2] Bennett seeks to complement structuralist approaches in the social sciences: "[T]he category of 'structure' is ultimately unable to give the force of things its due: a structure can act only negatively, as a constraint on human agency, or passively, as an enabling background or context for it."[3] She wants to strengthen the case for a positive view of *vibrant matter* and declares further: "To the roles of context, tool, and constraint (or background, resource, and limit) I will add the role of actant."[4] This term, adopted from the philosopher of science Bruno Latour, hints at things having a kind of activeness or vitality, which needs to be fleshed out and differentiated. Although the matters focused on here share the specific ability and purpose of making flows visible, they do so by intra-acting (as termed by Karen Barad) in an experimental setup-as-assemblage in various ways, and by adopting different roles. In historical case studies, it is nearly impossible to find out which elements formed ad hoc groupings additional to the experimental components of the setup mentioned in the publications and notes. However, it is possible to deduce a variety of ways to insert cuts through the moving liquid, revealing it being conceived as a Cartesian *a priori* space. The following sections discuss the various roles that markers adopt in fluid dynamics experimentation.

Being, embodying, and indicating the object of study

Marker material is often fabricated and tested in complicated processes. As soon as the visualization method or the marker itself is under explicit investigation, something becomes a host fluid that is given a different name. In the case of dyes, it might be called a *solvent*[5] or more generally the *working fluid*.[6] This indicates that marker materials are seen as *objects of study*.

On another level, substances may be investigated with regard to their role as *indicators* of a flow (the flow being the object of study) as opposed to being an *embodiment* of a flow. The former is the case in the 1989 research of Ismail H. Farhan and Francis G. MacCabee, who wanted to determine the locations of flow separation and boundary layer transitions around a high-speed train. In Fig. 1, we see that the air is enhanced with a whole front of tamed smoke streaks. The ideal line formation of the smoke streaks[7] hints at the air current. In my view, this is in contrast to the way that Georg Wellner (1846-1909), from the University of Brno, conducted his smoke studies on railway locomotives at the end of the nineteenth century. In addition to his pursuing other research agendas, he wanted to work out how trains or their chimneys should be constructed so as to minimize the inconvenience caused by smoke entering the open windows of the train. In contrast to approaches where the smoke streak-enhanced air front hits the train from the outside, in such a setup and due to the way the smoke of the funnel is conceived, it is what is studied in terms of behavior (not of its characteristics as a marker). Thus, here, it seems that the smoke is not only an *indicator* hinting

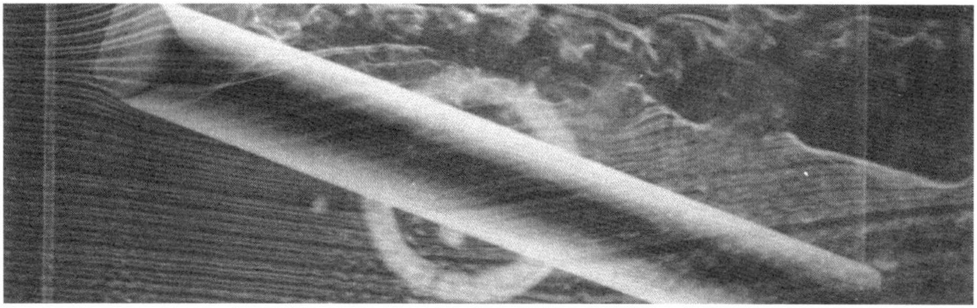

Fig. 1: Farhan and MacCabee, University of Technology Loughborough:
Flow pattern at 22° yaw angle using the smoke technique with an EMS train model, 1989.

at air, but it also *embodies* the untamed object of research. The full formation is developed in situ (exiting a funnel), while also existing in its optical and olfactory qualities. To put it another way, in this case the investigated object exhibits qualities of marker material while simultaneously sharing the location of its becoming.

Preparing the object of study

The German zoologist and hydrodynamic researcher, Friedrich Ahlborn (1858-1937), constructed his first water tunnel in 1901 and continued work on the concept until the 1930s. As many others, he visualized and observed the processes on the water surface of his trough.[8] To do this, he distributed fine-grained clubmoss spores on the surface of the water. Ahlborn's method of distributing the markers on the water provoked some debate. He was asked to what degree this was representative for processes in the interior of the water as the surface tension had to be taken into account. Ahlborn replied that the clubmoss spores strewn onto the water served to cancel out the tension on the water surface, as with any other dust.[9] In other words, in his view, the means to make flow visible simultaneously prepares the object of study in such a way that the natural surface can be seen as a conceptual section through the water volume. This is a second function lying in an unavoidable yet welcome side effect. The water surface has the big advantage of providing a privileged visual plane. It is obvious where to fix the camera's focus prior to the experiment.

Marking a two-dimensional (2D) space

However, Ahlborn could not use the clubmoss spores in the form he got them from the shops: "In earlier works I most often used clubmoss spores in the form they were commercially available. But when scattering them on the water surface it turned out that as soon as they touched the water many of the fine spore balls made a dancing movement and surrounded themselves with a free perimeter of 1-3 cm in diameter."[10] In this way, the spores self-organized the distances to each other. However, Ahlborn envisioned a denser distribution and wanted the particles to commit themselves more to the movement of the water instead of bouncing on their self-generated 'dancefloor'. I quote Ahlborn here at length, as he explains his troubleshooting process:

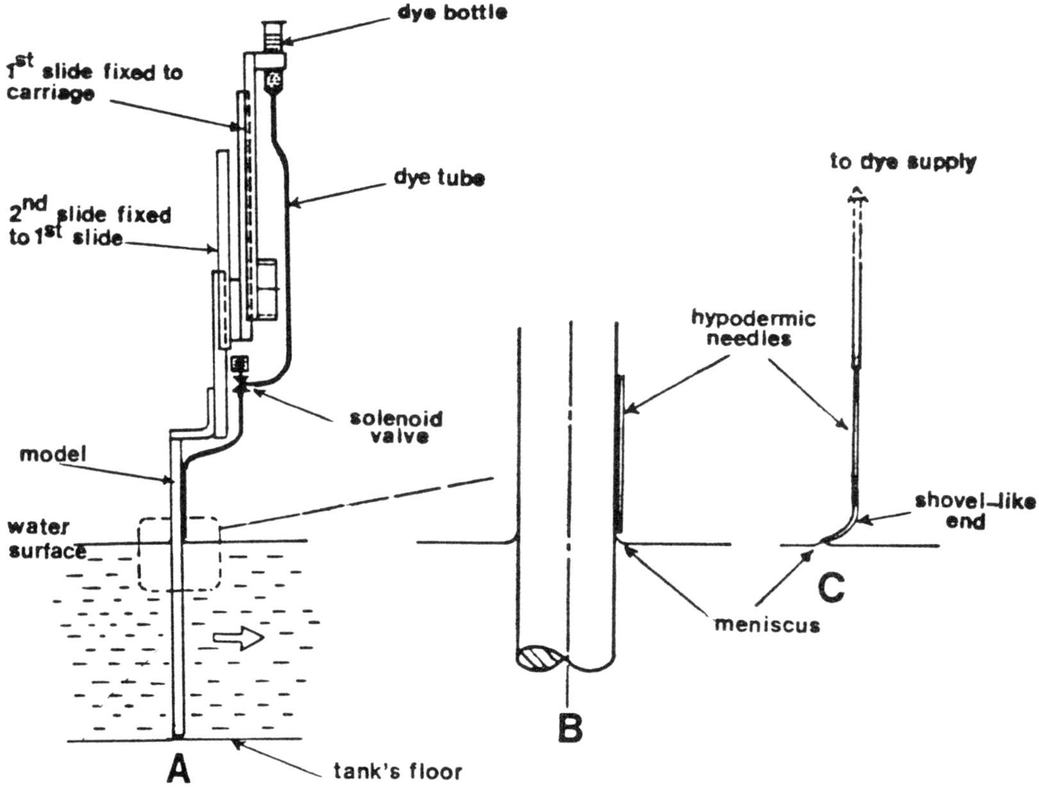

Fig. 2: Al-Khafaji and Gerrard, University of Manchester: Water surface flow visualization with dye, 1989 (showing how to introduce the marker).

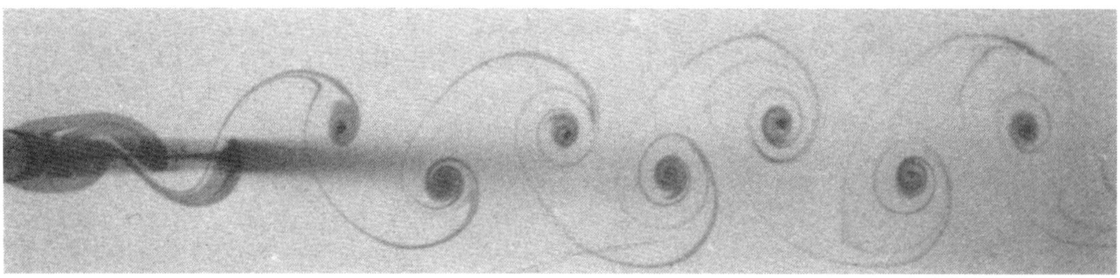

Fig. 3: Ibid. Introduced marker.

> *The material of the fine skin constituting the perimeter comes visibly from the clubmoss flour, and major efforts were necessary in order to generate a uniformly covered field of spores. Then the skin was obviously distributed equally over the whole surface, but not eliminated. By means of a careful extraction with alcohol it is possible to remove the skin-building material. It involved not insignificant amounts of yellowish essential oil, which detaches itself in the form of a dense emulsion as an addition to the water from the alcohol. Now, the well dried spores are no longer placed on the water by use of a shaker covered with a fine silk gauze, but by blowing them horizontally into the air with an atomizer nozzle about half a meter above the water so that they hang in the air in the finest distribution and when sinking down they distribute themselves very equally over the surface without aggregating into flakes. In this way one can easily attain a distribution of 4-6 spores on a square millimeter that proved to be especially suited for most purposes. Although the individual floating spores are hardly recognizable with the bare eye, with good lighting they nonetheless deliver very clear flow images, which with respect of resolution and sharpness surpass all former records of flow lines by far. However, the observation of these images has to be carried out with a magnifying glass or requires the prior magnification of the original photographs. As soon as the models are carried through the water enhanced with swimming or floating particles, they cause the movements upon which the resistance values depend and which become visible through the 'viewing bodies'.*[11]

However, these very delicate depictions, which must be inspected with a magnifier, are not suitable for reproduction in publications. Ahlborn therefore suggests first covering the water surface evenly with spores, as just described, and then, secondly, dividing this film with a small flat rod using gentle movements. This clusters the spores into small flakes. In order for them to appear as white flow lines in the photographs, it is important that they leave areas of the same size free between them. Contrast is a crucial issue. Ahlborn suggests blackening the background, and using shavings of aluminum bronze, which is optically very effective because of its white metallic luster. He preferred lycopodium, however, as the aluminum shavings tended to cluster and stick to the experimental models. In addition, when sinking to the ground, they made the background appear grey, and thus less rich, in contrast to the foreground phenomena.[12] From this we see that there is a clear sense of when such accumulations are welcome and when they are not. When the particles glue together, they are not always ready or in the right formation.

Supporting another marker

Scientists in fluid dynamics research pay close attention to ways that bring the marker and the object of study together. A. A. Al-Khafaji and John H. Gerrard from the University of Manchester developed a special apparatus (Fig. 2) for releasing markers onto the water (Fig. 3). The dark "dye solution which is prepared prior to the experiments is simply a mixture of a food colouring dye, tap water and a tiny drop of detergent. The amount of the detergent in the mixture is small enough for the dye solution not to spread significantly on the water surface when being introduced to it: without it the dye falls through the surface." [13]
Besides paying attention to the amounts of the ingredients, the way of suspending the dye is also a delicate issue. Here, the dye is "injected at a controlled rate onto a clean water surface from small diameter hypodermic needles rigidly and externally attached to the models." [14] These needles barely touch the water in order to release a precisely calculated amount of dye to make sure it remains on the surface.

Such caution would not be necessary with clubmoss spores. Even after they have been freed from their oil, they are perfect swimmers and can even carry other material, thus stabilizing an interface between different materials. In two handwritten notes from the year 1928 (kept in the Archive of the Deutsches Museum in Munich), Ahlborn describes an experimental setup to study the phenomena of boundary transition. These notes attest to another role given to the tiny lycopodium seeds. He seems to have divided the fluid of his open-surface test tunnel into three sections. All were covered with clubmoss spores, but one was also strewn with plaster flakes that provided sharply defined vectors. [15] If the related photographs could be found, we would hopefully better understand the qualitative difference between the lycopodium and the plaster. Ahlborn is clear about why he uses both kinds of particles: "[The s]urface is sprinkled with clubmoss and thereupon plaster in fine flakes that would sink quickly without clubmoss, while clubmoss can be blown away from it." [16]
In other words, clubmoss spores stay with the water only in the relative absence of air movement, whereas, in an ideal case, the marker particles would behave like the molecules of the studied fluid. This example makes clear that not every interaction with the markers results in their mimicking the behavior of the studied substance in

a detectible way. Thus, the task of the scientists could be reformulated as follows: How is it possible to turn a marker material into a visible water molecule, and, moreover, into one that can be traced so that it provides its location at any time?

Marking a three-dimensional (3D) space

The fact that the lycopodium particles cannot sink makes them unsuitable for so called three-dimensional (3D) studies, that is, the water volume (Fig. 4). What Ahlborn needed here were particles that sink very slowly, so that their vertical movement would not greatly disturb the water. However, they need to sink in order to occupy a 3D space in the interior of the water. Even if they could be turned into divers, there is a second reason "why small bodies of the magnitude of the clubmoss spores cannot be used in the interior of the water, and thus the current cannot be depicted in the same resolution and completeness as on the surface of the water."[17] The problem is that the spores cannot be forced into small groups in the interior of the water.

In order to make the inner movement of the water visible, Ahlborn used wooden chips from old oak or maple heartwood. These were freed from dust by filtering them several times and were cut down to a size of about one millimeter. This material was then washed and dried thoroughly and then preserved in a tightly closed container until use. Shortly before starting the experiment, the wood shavings were distributed in a slender wooden flume immersed slightly into the water and placed exactly above a slot built into the tunnel for lighting from below. In this way, the shavings were brought into a linear formation before they slowly began to sink and spread laterally. After a while, the whole depth was filled homogeneously with these floating bodies. Their sinking velocity was determined as negligible (2%) in comparison with the speed with which the test model was driven through the tank.[18]

Creating spatiotemporal sections of visibility

The counterpart of the careful positioning of the markers is to be found in an equivalent lighting strategy, which tries to prevent the light spreading in all directions by using mirrors to deflect the light or shield it, except for a narrow aperture. This allows the wood shavings to be illuminated intensively "in the image plane".[19] This is symptomatic. Again and again in the technical literature, we encounter solutions that define aisles and sections in a continuum in order to guarantee visibility. Some of the major actants (markers, background, light, and exposure time) are meant to work toward an extended two-dimensionality or a flattened three-dimensionality.

For example, in particle-image velocimetry (PIV), "[p]articles in the fluid are illuminated by a sheet of light that is pulsed. The particles scatter light into a photographic lens located at 90° to the sheet, so that its in-focus object plane coincides with the illuminated slice of fluid."[20] Lighting is also used in order to address only a part of the whole experimental setup. Where tracer material lies in the dark, it does not make anything visible for the recording; on the contrary it should not hinder the view of the defined and illuminated area.

An experiment to determine the mechanics of an insect's wings when flying gives an example of a combined and concerted situation that is also temporally and spatially (quasi geometrically) constricted. The experiment was conducted at the University of Tokyo in the 1980s. In the 'restrained fly' variant, the insect was carefully fixed by its belly with an adhesive on top of a string, whereby the scientists paid attention to letting its legs free so that it can position them as in a real flight. Now imagine the groovy, multimodal experience that this poor dragonfly had to endure. Smoke streaks, smelling of the paraffin with which they were made, floated toward it in either horizontal or vertical formation. After the streak flow had passed the insect by some centimeters, flash lights were added for multiple exposures. The scientists explain that: "The tunnel was designed and built to generate many smoke lines arranged in a cruciform and to be capable of three-dimensional observation of the flow."[21] Interestingly, the depictions accompanying the article only show images with either a vertical (Fig. 5) or a horizontal alignment (Fig. 6a-c). In this way, overlaps due to the 2D photographic projection are avoided as much as possible. The scientists studied the lift and thrust of the flight. In Fig. 6b, we see what they called shed vortices, whereas in Fig. 6c we see trailing vortices. Here, light co-defines local and temporal slices in cooperation with visible particles.

Another kind of selectivity is gained when well-dosed light triggers temporal blinking between visible and invisible states of certain markers, allowing the camera to profit from this in actu discretization of its motif. This punctual, timed visibility can be created by photochromic flow visualization (PFV) techniques, the principle of which was invented in 1967 by A. T. Popovich and Richard L. Hummel for ethyl alcohol.[22] In the absence of short wavelength light, photochromic dyes are colorless when in solution in many liquids, but "may be selectively activated by an ultraviolet light source to produce a bright color". The dye used in the experiments of Cecilia Mansson, Jesse Maddren and Ekkehard P. Marschall is TNSBP: "TNSBP is soluble in most aliphatic and aromatic solvents, and many other nonpolar liquids, it activates in less than 3 microseconds, and it will remain in the colored state up to and in excess of thirty seconds depending on temperatures. It … will deactivate after a certain period of time and return to the colorless state."[23] Other researchers used substances where they did not have to wait until visibility faded again by exploiting the substance's ability to change appearance when brought in contact with pulse laser ultraviolet radiation. V. N. Yourechko and colleagues used a substance from the class of indoline spiropyranes: "TMINBPS has two steady forms: colorless or pale yellow and deep blue or purple when activated. The reverse reaction is activated by green light or heat."[24] This is a kind of reversible dye technique where the visibility of the patterns can be erased. Apparatus-related visibility is created with the sparing use of tracers and a very partial lighting. There is no excessive use, but rather deliberate restrictions of various kinds to enable the necessary distinctions.

Making visible – but for whom or what?

At the very latest when we notice that most of the experiments take place in shaded rooms or even in perfect darkness "because of the convenience of photographing"[25], interrupted only by a short flash of light, a fired strobe light,[26] or a series of laser pulses, we have to ask ourselves for whom this visibility is generated. Although sunlight is excluded, windows are a necessary prerequisite for these experiments. They are made for inspection, lighting or recording and denominate those parts of the tunnels that are made of glass, plexiglas or acrylic. Ahlborn also mentions that, in these 3D studies, the lateral space next to the viewing window (most often occupied by the camera) should remain clear and the shavings should not be placed too close to the resistance object to prevent blocking the view (Durchsicht) and the light (Durchleuchtung).

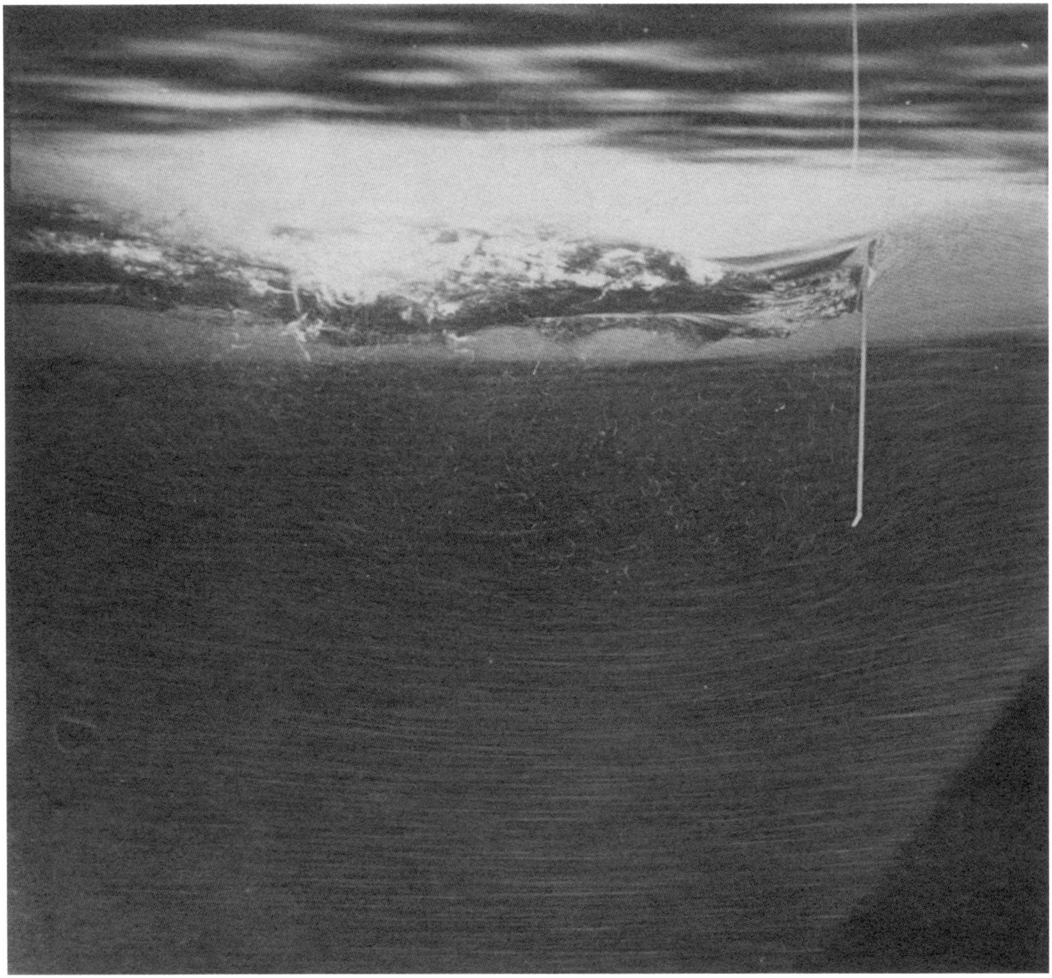

Fig. 4: Ahlborn: Recording from the inside of the water, with a view toward the water surface and the immersed plate.

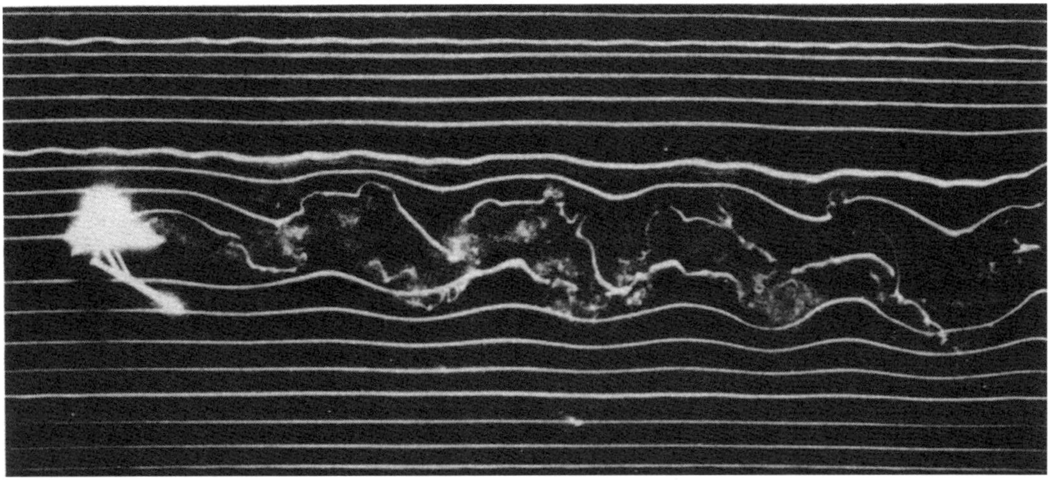

Fig. 5: Watanabe et al., University of Tokyo: Dragonfly's restrained flight, 1986.
Vertically-aligned smoke streaks, profile view.

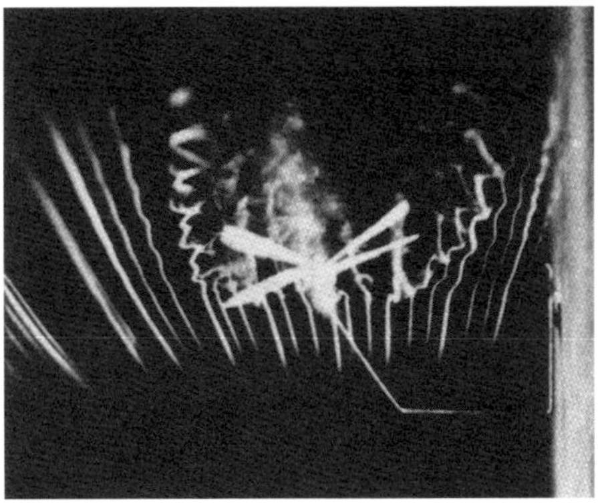

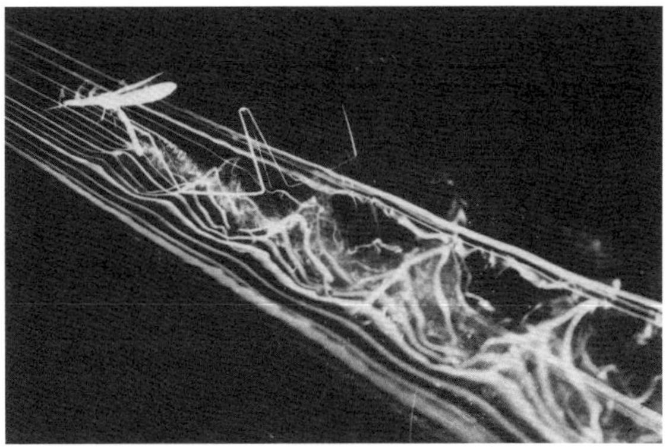

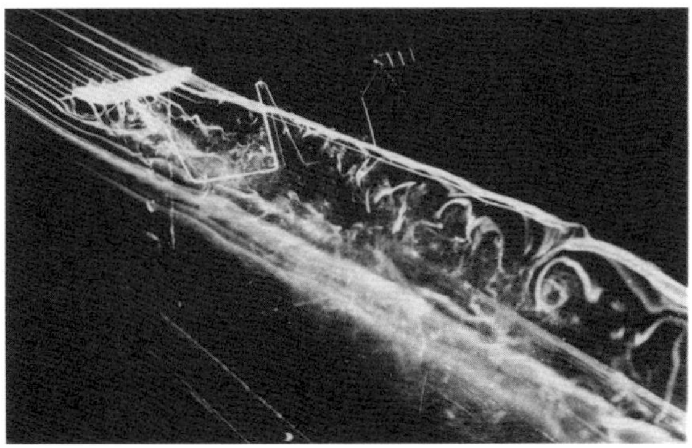

Fig. 6a-c: Ibid. Horizontally-aligned smoke streaks:
a) frontal view;
b) shed vortices;
c) trailing vortices.

When providing visual clues, there is a very fine line between too little or too much, thus obstructing the view. The above-mentioned, clearly defined, linear formation of the particles is nothing other than a method for adjusting the right portion and for artificially generating a section.

When reading the review on flow visualization techniques by the engineer Thomas J. Mueller, it becomes obvious that the lighting apparatuses play a crucial role in the experimental setup. The engineer Wolfgang Merzkirch goes so far as to group the field of flow visualization in the domain of photonics. Photonics embraces all methods wherein the information is produced, transported, and recorded by means of photons.[27] Very soon after some trials to inspect the events visually with one's eyes, the researchers focused on adapting the situation for photographic or filmic recordings or other measurements. Perhaps not surprisingly, a major discourse around hydro and aerodynamics from around 1900 onwards is concerned with issues such as light and contrast.

In addition, researchers have worked hard to ensure that the light has a reliable partner medium: the smoke should be dense enough, and the introduced particles–if not themselves flames or fluorescent–should sparkle or gleam. This shows that the materials introduced to the wind and water tunnels were selected according to their luminosity because they had to communicate with the recording devices.

The third agent in play is the background, guaranteeing a clear contrast to the tracer materials. Often it is painted with a dark lusterless wash or black velvet is used. In the reverse case–when darker dyes are used–it is made white.

Providing data for further visualizations

Ahlborn succeeded in conducting a few experiments with markers on the water surface on a larger scale in the research laboratory of Norddeutscher Lloyd in Bremerhaven. Whereas he used plates of about ten centimeters as obstacles in his water trough in Hamburg, here a steel plate of one square meter–five millimeters thick–was half immersed in the water. Due to the larger dimensions, the camera was also positioned at a greater distance, about three meters above water level and directly orthogonal to the plate. As a result, the resistance object appears as a straight line in the photographs (Fig. 7). This greater distance necessitated other markers. He used so-called 'confetti', made from circular punched-out paper shavings five millimeters in diameter, which did not sink. In

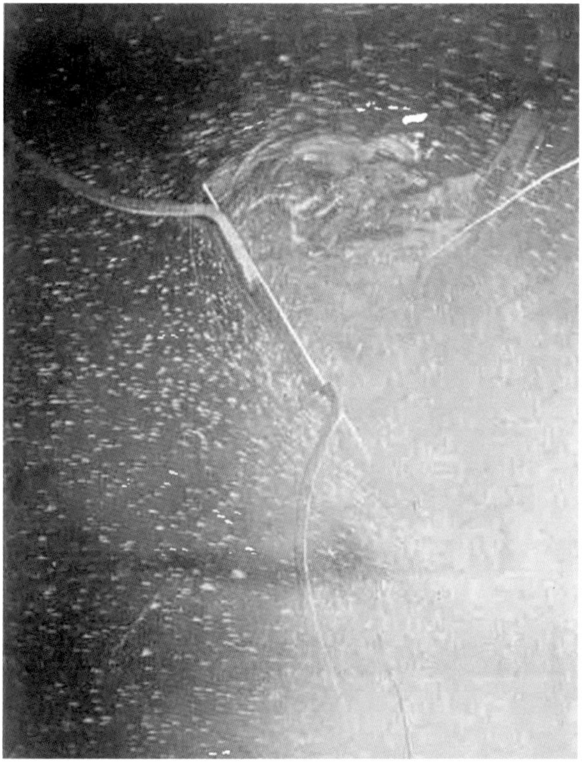

Fig. 7: Ahlborn: Experiment with paper shavings in the Norddeutschen Lloyd laboratory in Bremerhaven, ca. 1902.

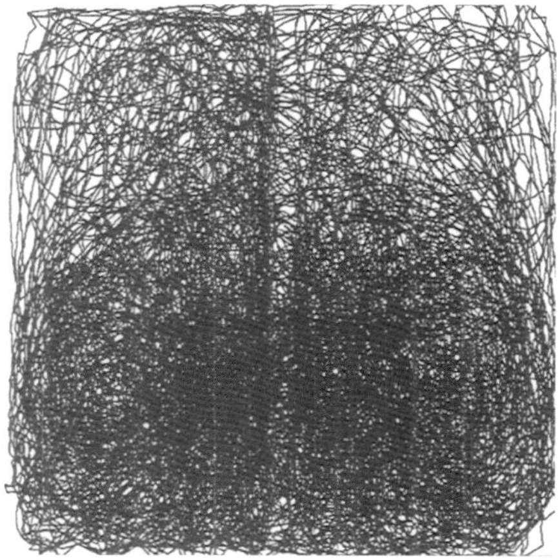

Fig. 8: Wittmer et al., Université de Lorraine, Nancy: Retraced trajectory of a particle in a stirred tank over a period of 27 minutes, 1995.

one instance, Ahlborn used tetraedrical spores, whereas in another he used flat paper confetti more than a hundred times larger, although both were used to represent the same water. Due to the bigger size of the marker, the flow lines are equally well visible. However, due to the camera's distance, given the same period of exposure, their trajectory is shorter than that of the tiny clubmoss spores (0.03 mm) in the small trough.[28] In other words, the traces of the markers have different proportions in the photographs. This must be taken into account when analyzing the photographs and making further drawings from them.

Further thought deserves the following remarkable statement by Stephan Wittmer and colleagues from the École Nationale Supérieure des Industries Chimiques in Nancy: "a certain number of tracer-particles are introduced in the flow and visualized in order to get local positions".[29] Here, markers are introduced, recorded, and further visualized. Although it could be that by visualization they primarily meant the retrospective extraction of the 3D position of one particle in a tank from the film recordings (Fig. 8), it is also valid for the experimental situation itself. Using visual things and making things visible are two different issues and it is interesting to investigate the tension between them.

1 Merzkirch, Wolfgang: *Why making use of flow visualization?,* in: *Journal of Flow Visualization & Image Processing*, Vol. 19, No. 2, 2012, pp. 161-177; here: p. 162.
2 Bennett, Jane: *Vibrant Matter. A political ecology of things*, Duke University Press, 2010, p. viii.
3 Ibid., p. 29
4 Ibid., p. 39
5 Mansson, C., Maddren, J., and Marschall, E. P.: *Photochromic Flow Visualization in Non-Isothermal Liquid-Liquid Two-Phase Flow*, in: Tanida, Y. and Miyashiro, H. (ed.): *Flow Visualization VI*, Springer, 1992, pp. 550-554; here: p. 551.
6 Akiyama, M., Suzuki, M., and Hishiwaki, I.: *Transport Phenomena of Ventilating Flows in a Rectangular Room*, in: Asanuma, Tsuyoshi (ed.): *Flow Visualization I*, Hemisphere Publishing Corp. 1979, pp. 135-141; here: p. 136. Yourechko, V. N., Ryazantsev, Y. S., and Ilminsky, V. N.: *The Method of Photochromic Visualization in Liquid Flows*, in: Rezníček, R. (ed.): *Flow Visualization V*, Hemisphere Publishing Corp. 1989, pp. 65-71.
7 Farhan, I. H. and Maccabee, F. G.: *Wind Tunnel Flow Visualization Tests on an Electromagnetic Suspension Train Model*, in: Rezníček, R. (ed.): *Flow Visualization V*, Hemisphere Publishing Corp. 1989, pp. 615-622.
8 For an overview of his experimental setup, cf. Hinterwaldner, Inge: *Parallel Lines as Tools for Making Turbulence Visible*, in: *Representations*, Vol. 124, No. 1, 2013, pp. 1-42.
9 Ahlborn, Friedrich: *Über den Mechanismus des hydrodynamischen Widerstandes*, Friedrichsen 1902, p. 29.
10 Ahlborn, F.: *Versuchsabteilung der Flugzeugmeisterei*, pp. 3-5; here: p. 4 and supplement. Undated typescript, Deutsches Museum Archive (DMA) NL Ahlborn k72. [transl. by the author]

11 Ibid., p. 39
12 Ahlborn, F.: *Die photographische Strömungsanalyse. Versuchseinrichtungen und Verfahren*, undated (after 1917), pp. 1-7, here: pp. 1-4 (after 1917). Typescript, DMA NL Ahlborn b-88.
13 Al-Khafaji, A. A. and Gerrard, J. H.: *Water Surface Flow Visualization of Bluff Body Wakes*, in: Rezníček, R. (ed.): *Flow Visualization V*, Hemisphere Publishing Corp. 1989, pp. 405-413; here: p. 408.
14 Ibid.
15 Ahlborn, F.: *3 Aufnahmen mit sehr dicken Zylinder-Oel Kraftlinien. Umströmung eines Durrplattenversuchs. Eine Lampe*, January 27, 1928. Manually written note, DMA NL Ahlborn 035.
16 Ahlborn, Friedrich: untitled, January 1928. Manually written note and sketch of the experimental setup, DMA NL Ahlborn 035. [transl. by the author]
17 Ahlborn, (as fn 12), p. 6. [transl. by the author]
18 Ibid., p. 5. Ahlborn, F.: *Über den Mechanismus des hydrodynamischen Widerstandes*, in: Naturwissenschaftlicher Verein in Hamburg (ed.): *Abhandlungen aus dem Gebiete der Naturwissenschaften*, Vol. 17, Friedrichsen 1902, pp. 29-30.
19 Ahlborn, F.: *Hydrodynamische Experimentaluntersuchungen*, in: *Jahrbuch der Schiffsbautechnischen Gesellschaft*, 1904, pp. 417-453; here: p. 421. [transl. by the author]
20 Adrian, Ronald J.: *Particle-Imaging Techniques for Experimental Fluid Mechanics*, in: *Annual Review of Fluid Mechanics*, Vol. 23, 1991, pp. 261-304; here: p. 263.
21 Watanabe, Isao, Azuma, Akira, and Watanabe, Tadaaki: *Wake Vortices of Flying Dragonflies*, in: Véret, Claude (ed.): *Flow Visualization IV*, Springer, 1987, pp. 821-826; here: p. 821.
22 Popovich, A. T. and Hummel, Richard L.: *A new method for non-disturbing turbulent flow measurements very close to a wall*, in: *Chemical Engineering Science*, Vol. 22, No. 1, 1967, pp. 21-25.
23 Mansson, C., Maddren, J., and Marschall, E. P.: *Photochromic Flow Visualization in Non-Isothermal Liquid-Liquid Two-Phase Flow*, in: Tanida, Yoshimichi, and Miyashiro, Hiroshi (ed.): *Flow Visualization VI*, Springer 1992, pp. 550-554; here: p. 551.
24 Yourechko, V. N., Ryazantsev, Yu. S., and Ilminsky, V. N.: *The Method of Photochromic Visualization in Liquid Flows*, in: Rezníček, R. (ed.): *Flow Visualization V*, Hemisphere Publishing Corp. 1989, pp. 65-71; here: p. 66.
25 Hirayama, T., Nagai, M., and Ueno, I.: *Visualization of Artificial Transient Water Wave*, in: Asanuma, Tsuyoshi (ed.): *Flow Visualization I*, Washington: Hemisphere Publishing Corp. 1979, pp. 123-128; here: p. 124.
26 Yamada, Hideo: *Use of smoke wire technique in measuring velocity profiles of oscillating laminar air flows*, in: Asanuma, Tsuyoshi (ed.): *Flow Visualization I*, Hemisphere Publishing Corp. 1979, pp. 265-270; here: p. 265.
27 Merzkirch, W.: *Flow Visualization Research in Western Europe*, in: Asanuma, Tsuyoshi (ed.): *Flow Visualization I*, Hemisphere Publishing Corp. 1979, pp. 29-36; here: p. 29.
28 Ahlborn, (as fn 9), p. 23.
29 Wittmer, S., Vivier, H., Falk, L., and Villermaux, J.: *Three-dimensional long-term particle tracking in a stirred tank*, in: Crowder, James P. (ed.): *Flow Visualization VII*, Begell House, Inc. 1995, pp. 628-633; here: p. 628.

Sources of the Figures:

Fig. 4: Deutsches Museum, München, Archiv, CD_62752
Fig. 7: Ahlborn, Friedrich: Über den Mechanismus des hydrodynamischen Widerstandes, Friedrichsen 1902, p. 23, Plate XI, Fig. 50.
Fig. 8: Laboratoire Réactions et Génie des Procédés, UPR 3349 CNRS, Université de Lorraine, 1 rue Grandville, BP 20451, 54001 Nancy Cedex, France. With kind permission of Pascal Pitiot, Stephan Wittmer, Hervé Vivier, Laurent Falk, and Jacques Villermaux.

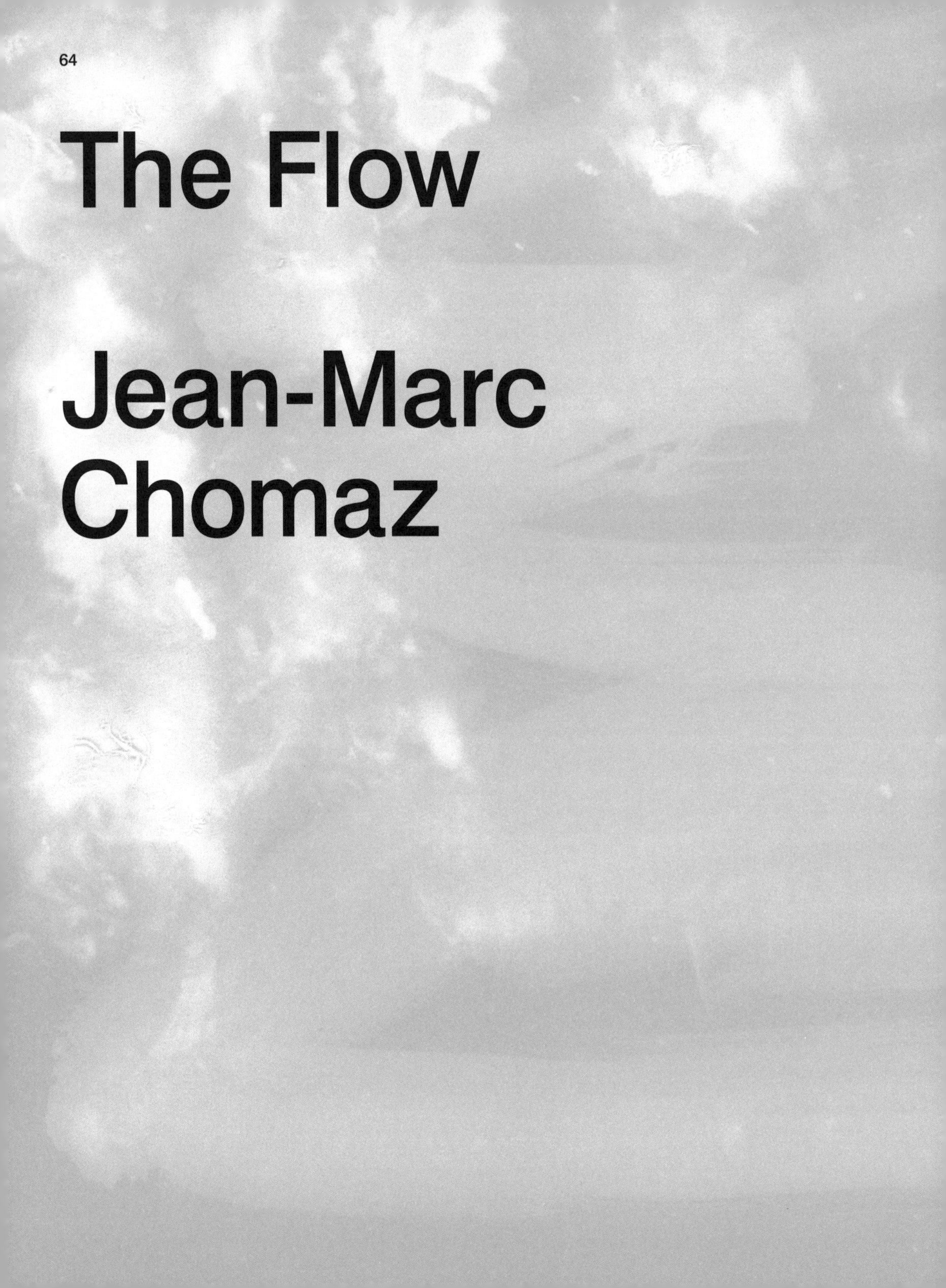

The Flow

Jean-Marc Chomaz

Science views the universe as a dynamic system, from the cosmic to human to subatomic scale; as a trajectory rushing through the ages; as a gigantic flow which shapes all structures and textures, visible or invisible to the naked eye. Following Arnold, we will use the term *the Flow* to refer to Anosov's flow, which is defined as the bundle of all trajectories of all the restless particles in our universe, and of all the degrees of freedom in the emptiness of space. The Flow is the effective form science is bound to invent; a form composed of motion, matter, and trajectories. Our current perspective, that of an expanded reality made more tangible by technology, is only a snapshot: an image of the Flow at a particular instant that we perceive of as *our age*. Our observations of the depth of the sky do not capture an ancient point in the age of the universe, but rather intercept it on its race to the future: particles, photons, and neutrinos, emitted during an ancient transformation from one energy form to another, propagating in continuous interaction with the rest of the universe. In what we perceive of as *the present*, these particles and processes are transformed by our extended senses into images that the electrons in our brains can make sense of.

The Flow is tremendously complex. Even if each individual particle obeys elementary deterministic rules, collective forms emerge at all scales in time and space transcending the rules individual particles are submitted to. Our senses interpret these collective structures as mental images or projections: n-dimensional objects or concepts that are a pale perception of the complexity of forms which shape the spatio-temporal Flow. However, our minds unconsciously detect these complex morphodynamics when they occur at our scale of time and space–for example, the mystery of clouds arises from their apparently still yet ever-changing form. Evelina Domnitch and Dmitry Gelfand's *Camera Lucida* installation has succeeded in manifesting the Flow in a technological setting. It consists of a spherical tank inside which sonoluminescence is generated via the interaction of three transducers to produce intense ultrasonic beams. In doing so, the installation creates evanescent boreal auroras and evokes a universe that seems to emanate from our own thoughts.

Science continuously proposes partial representations of the Flow, at cosmic, geologic, or subatomic scales of time and space, which are aimed at describing morphodynamics compatible with present observations. These models project a form where matter, space, and time are in motion, allowing our imagination to screen and insinuate scenarios with the feeling of embracing a larger view of the Flow than our minds may ever be able to grasp.

In an objective sense, science gives physical existence to the Flow by formulating simple rules of symmetry, conservation, and causality. Mathematics can therefore be regarded as a language that represents the Flow in a mental form, revealing unforeseen principles and phrases that, when formulated, allow for definitive new tests to challenge accepted measurements. Invariably, newly formulated observations will question the Flow model, and scientists will be compelled to discard their old constructions and build a new language: a novel mathematical grammar with new physical verbs that will generate a new mental projection of the Flow to explain both past and present data. All of the hypotheses – however poetical they may be – that the reinvented language may generate should be patiently written, and their effects carefully tested against the observations in the expectation to discover an inconsistence, a dissonance in the real that will bring us once again roll the stone of knowledge from the base to the top of the mountain, free and happy to reformulate the world.

Mixing and segregation are two processes that shape the Flow at all scales of time and space, and explain the textures of the instantaneous images our minds project. In the present model of the cosmic Flow, the initial uniformly composed dense cloud of particles expand into space, and then concentrate under the action of gravity from vanishing fluctuations. The induced segregation of mass and momentum creates primitive, massive stars that transform primordial matter into heavier atoms, which rapidly blow outward into the present dusty universe; where, from gigantic dark clouds, galaxies arise. This ancient period of the Flow is presently tested with computers, which implement the language of mathematics to automatically create new generations of phrases called *results*, and to compare predictions of the texture of the 3 Kelvin Background Radiation with present observations. This proto-universe model is certainly different from the real Flow, but confronting the model with the present should enable the validation of assumptions about these remote ages and the explanation of the current distribution of galactic clusters. It is particularly conjectural and fascinating, because the remnants of these primordial epochs are hidden from us, behind the blossoms of particles emitted during explosions of these primordial-stars, explosions nearly synchronized since occurring at approximately the same instant after the Big start of the Flow.

Inside actual galaxies, stars are constantly being generated by the high energy interaction of molecular clouds, which create shockwaves that in turn produce fluctuations and gravitational collapse. These galactic clouds are made up of dust and molecules; they diffuse the light of the stars behind or within them, creating intricate patterns which incite our imagination to see familiar shapes or animals. They resemble atmospheric clouds, since they obey a similar law of mixing and segregation to preserve rotational invariance resulting in the conservation of angular momentum. Coherent structures appear to physically manifest these laws, while at the same time, mixing occurs; cloud plumes are chaotically transported, folded, and stretched at spatial and temporal scales compared to which humanity's existence is but a mere blink. The structures eventually condense into proto-stars wherein the laws of conservation of angular momentum and energy of agitation are balanced by gravitational collapse. Proto-stars transfer their momentum to dusty accretion disks from which, by a process still hidden from us, planets emerge – like a giant whirlpool concentrating the surrounding clouds into proto-planets. Again matter segregates, under the simultaneous action of gravity and rotation which defines planetary composition. Volatile elements then combine to form ethereal, yet superficial layers called oceans or atmospheres, the compositions of which vary amazingly when we consider our immediate surroundings: the carbon dioxide- and nitrogen-dominated Venus or the hydrogen- and helium-rich Jupiter. Thermal forcing of the atmosphere by the Sun or by internal heat sources induce convective motion that forms vortices to conserve angular momentum. These vortices structure the atmosphere, transporting heat and chemicals around but in isolation from the planetary core, while at the same time promoting the mixing of surrounding fluids via stirring and folding at their periphery. Minor constituents undergo phase changes amid liquids, solids, or gases that render both the folds and core visible: this creates clouds which can appear white when made of water (on Venus, Earth, and Jupiter, for instance) and multicolored when made of sulphuric acid, ammonia ice, or ammonium hydrosulfide). Among all of these cloudy maelstroms, the Great Red Spot of Jupiter, first been observed by Galileo, is clearly the most recognizable. The mechanisms responsible for its existence and color are still debated, but what is known is that The Great Red Spot is a vortex preserved from dispersion by the conservation of its angular momentum, which creates a barrier against mixing with the external fluid and secures its colorful segregation. Our project, *Luminiferous Drift*, in collaboration with Evelina Domnitch and Dmitry Gelfand, aims to examine a recently discovered stable coherent structure at the south pole of Saturn that strongly resembles a white hexagon. It is believed to have been created by destabilization of the shear present at the edge of the circumpolar vortex, shear that acts as a barrier against mixing shapping the edges of the hexagone. Our installation will

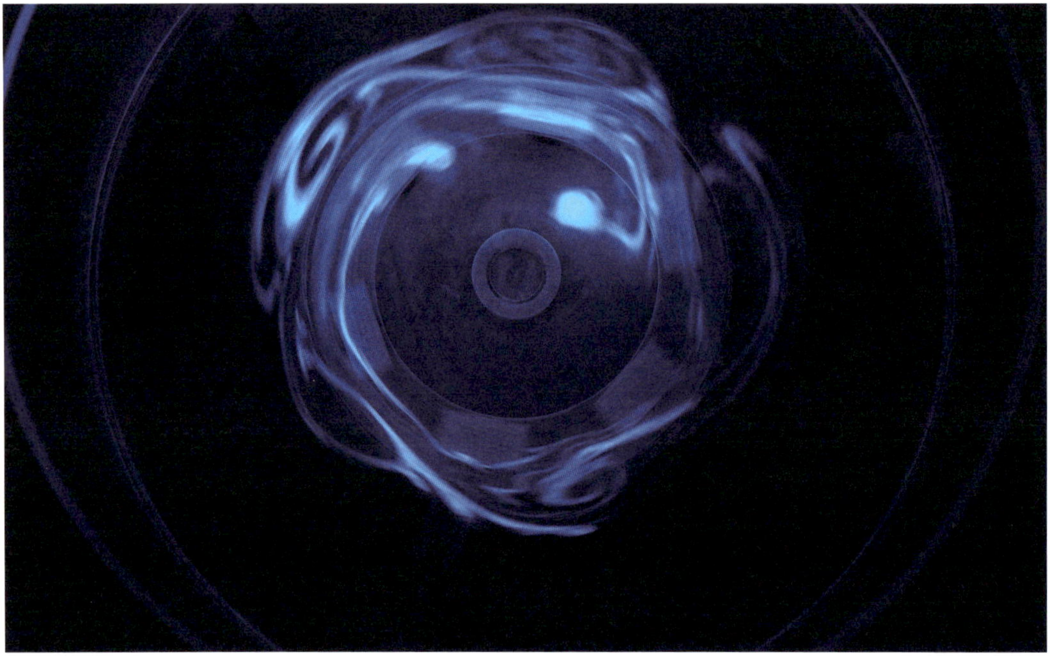

Fig. 1: *Luminiferous Drift* Installation, Evelina Domnitch, Dmitry Gelfand, and Jean-Marc Chomaz as created for the exhibition Wetware, curated by Jens Hauser and David Familian.

mechanically reproduce the vortex and the shear, and via the segregation of bioluminescent proto-cells, will also explore the potential future of Saturn's vortex from the perspective of climate change immanence in the universe.

On Earth, cloud dynamics are better understood, and also involve a similar mechanism of mixture and segregation. The processes from which they originate determine their texture. As we all know, Earth's white clouds are composed of water droplets or water ice crystals that condensate when an air mass is cooled down, generally by upward movement due its own buoyancy, topographical features such as mountains, or colder air masses flowing beneath it. It may also cool down due to heat loss or mixture with colder air. A summertime thermal plume, rising above a cotton field, produces a clear-sky cumulus cloud, and soil moisture condenses out of the rising hot streams of thin air.

Stormy weather is generated by a similar but far more dramatic process. When hot, humid, saturated air, often of tropical origin, is pushed upward by cold northern air, a potentially catastrophic sequence of events occurs. The moisture condenses when the depressurized air, moving upward and releasing its latent heat, warms the air and accelerates its ascent; at the same time, rain, rain is coming down, all over the intruding lower layer of fresh air cooling it more. This process progresses even faster, consequently destabilizing the entire region. A mountain standing against the wind will also force air to rise, which can then condense into layers of lenticular clouds when the humidity and vertical displacement are large enough. Amazingly, these topography-based clouds appear stationary, even though they originate from the fast displacement of air as it continuously enters and exits the clouds. Often, the air will oscillate up and down as it moves away from the mountain, producing a spatial echo of its shape. On occasion, clouds can become trapped or twisted into vortices because of mountains or other topographic features, producing the kinds of vortex streets that impressed the first cosmonauts by the arabesques they traced over our paper planet.

These topographic effects, when associated with the trade winds, explain rainfall on the windward sides of tropical islands and many other microclimates, as well as the drier and sunnier weather on the west cost of Leeward islands. Although topographic clouds are periodic in terms of space, other types of cloud patterns are periodic in terms of time. Sea breezes, for example, are diurnally evolving cloud patterns that commence when cold air from the sea moves inland during the day, upwardly displacing warmer air from the land, and generating a band of clouds that hug the shoreline until nightfall, when the wind reverses direction as the land cools off faster than the sea. Mountain breezes are similarly heated by the Sun on the *adret* (or sun-receiving) side of a mountain, only to then be cooled in the shade of the *ubac* (or sun-deprived) side as they spill down into the valley.

All clouds have distinctive shapes that arise from the very process of their generation. Plumes and cumulus borders are rounded by horizontal vortices that mix the cloudy air with the drier surrounding air, evaporating the droplets. High altitude stratus clouds are torn apart by chaotic motions induced by large-scale horizontal vortices. Lenticular clouds or caps may either appear extraordinarily smooth like flying saucers, or more paste-like when the wind is stronger or the stratification becomes weaker.

Clouds can also form large-scale patterns resembling planetary circulation, like in the case of a tropical convection band that is surrounded by a clear, dry subsidence zone. Then, at mid latitude, the scenic band of rainy depression, coming from the Rossby wave undulating on the northern potential vorticity, appears as a distinguishable vortex in the sky. Again, these patterns are made visible by the transportation of moisture, either vertically or horizontally, via vortices at all scales of space and time. Because of the conservation of angular momentum, the gyroscopic effect prevents fluid in the vortex from mixing, in contrast to the strongly mixing outside vortices, both of which are impelled by the same ever-present stretching and folding of space.

The formation of clouds, however, is not so simple, since without hydrophilic dust–called condensation nuclei–no cloud could form, even if the air was oversaturated. The energy needed to form even the first infinitesimal drop is much too large to be incited by random encounters of agitated water molecules. Water does not condense into droplets spontaneously, but rather into small hydrophilic particles which allow them to take on a finite size, larger than that at which they would evanesce. But when the rain falls it washes out these particles and clears up the sky–and indeed, it looks different, deeper, after the rain. The condensation nuclei must then be regenerated. Amazingly, these nuclei mainly come from the ocean, as salt particles from dried sea spray or as biological particles from the marine water, both of which are formed directly at the singular cusp of wave breaking against the wind, as well as through an explosion of micro air droplets formerly trapped beneath the broken waves. A similar chaotic process that moves masses of moisture around the planet, transports and spreads sea-born nuclei throughout the atmosphere and across the surface of the Earth. The nature, density, and size of the condensation nuclei influence the *albedo*, or optical density and reflectivity of the clouds' texture. Amazingly the atmosphere is permanently seeded and inhabited by plankton, which deepens the relationship between clouds and the ocean. In our project, *Luminiferous Drift*, we use synthetic biology to produce luciferine protein encapsulated in vesicles to reproduce the luminescence of some of the Earth's Phytoplankton. These vesicles, in the scenario of the project's installation, are strongly related to life at its origin, to the air encompassing oceans, and to extraterrestrial life. The barely *luminous* hexagon that starts vacillating in the installation is a reference and probe to the curiosity of Saturn's current vortices.

When humans started burning their fossil past in order to power their industrial future, they not only released greenhouse gasses but also anthropogenic condensation nuclei such as sulfate, nitrate, and ammonium. The first types warmed the atmosphere, while the second types increased the density of the clouds and, by reflecting solar radiation back into space, cooled the atmosphere. Simply put, man-made clouds did somewhat compensate for global warming by hiding the

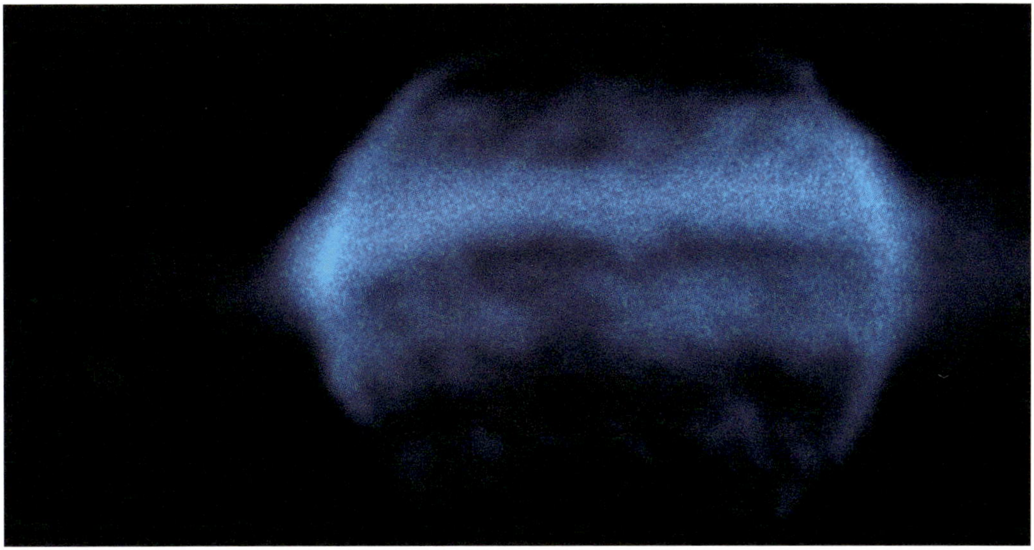

Fig. 2: *Exoplanet*, Salon des Réalités Nouvelles, Paris 2015. The installation Exoplanet is an intimate inverted cosmos inhabited by the bioluminescence of living phytoplankton Pyrocistis Noctiluca. Pale blue pulses draw a vortex, an evanescence of spirals and stripes. Sparks are produced by the oxidation of the luciferin, a protein photosynthesized during the day by living cells.

greenhouse effect. However, because man-made clouds did produce acid rain, the United Nations succeeded in regulating their emission while unraveling the imminent greenhouse gas catastrophe. This catastrophe is in motion within the Flow; its dynamics are hard to predict, as they are leading to an unprecedented balance of the biosphere, atmosphere, ocean and *homo-economicus-sphere*.

Climate engineering, another of Pandora's temptations, would be an energy industry-led attempt to control the climate by creating planet-size clouds whose purpose is to introduce acidic nuclei into the high atmosphere. Is it reasonable to control such a large part of the Flow, so gigantic when compared to human scale? Is it ethical? Or shall humans just burn their wax wings, imagining themselves as rivals to the Sun, not leaving enough hope to flow with their divine ambitions?

1 Vladimir Igorevitch, Arnold: *Chapitres supplémentaires de la théorie des équations différentielles ordinaires* (Translated from French by Djilali Embarek), Editions Mir Moscou 1978; in English: Vladimir Igorevitch, Arnold: *Lectures on partial differential equations*, Springer 2004

Winding the Vacuum

Evelina Domnitch and Dmitry Gelfand

According to our cosmist predecessors, humanity's weightless presence beyond Earth's atmosphere was not merely a fanciful possibility, it was considered an unequivocal vector of terrestrial evolution as well as a necessary perceptual leap for contextualizing Earth within a larger cosmic framework. Our pursuit of weightlessness and the cosmic vacuum, where it is most commonly experienced, was initially fulfilled through an artwork called *Sonolevitation* (2007). A high-pitched sound rises from a transducer and bounces back upon itself from a flat surface a few centimeters above. This acoustic *standing* wave evenly restructures the aerial fluid into zones of augmented sonic pressure interspersed with semi-vacuous holes. Hovering in these holes, known as *antinodes*, are leaves of gold that vibrate and spin almost without friction as they modulate the amplitude and frequency of the standing wave that levitates them. The slightest change in a

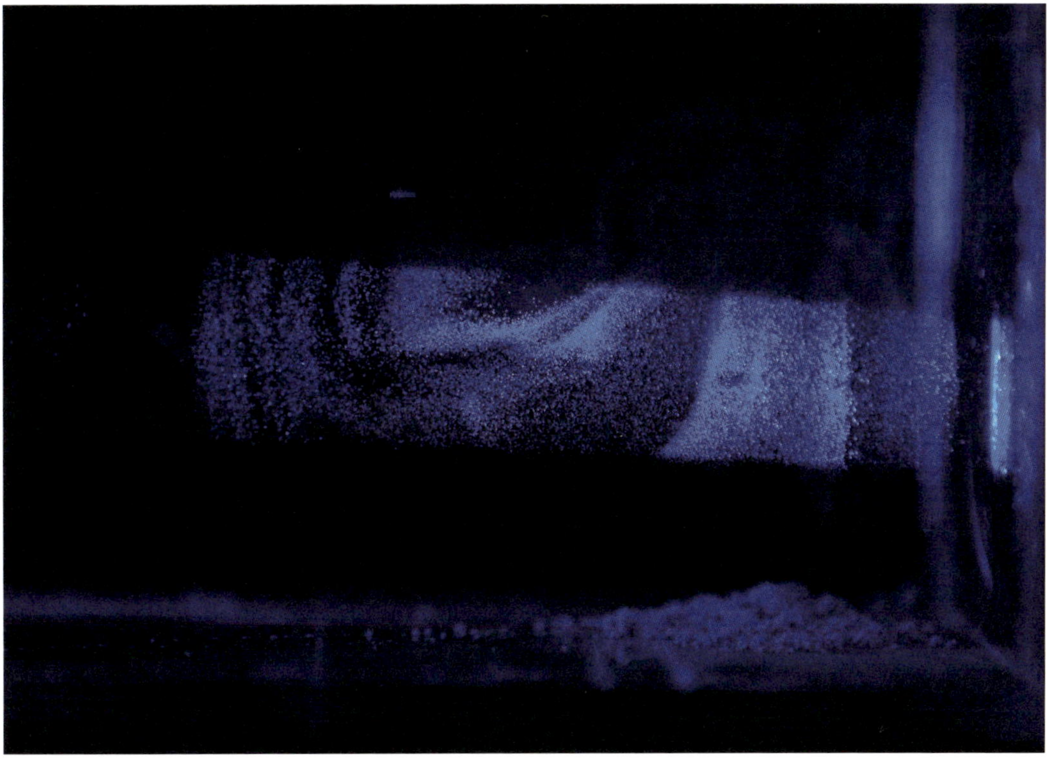

Fig. 1: Vortical formation of optically-levitated diamond dust in Photonic Wind.

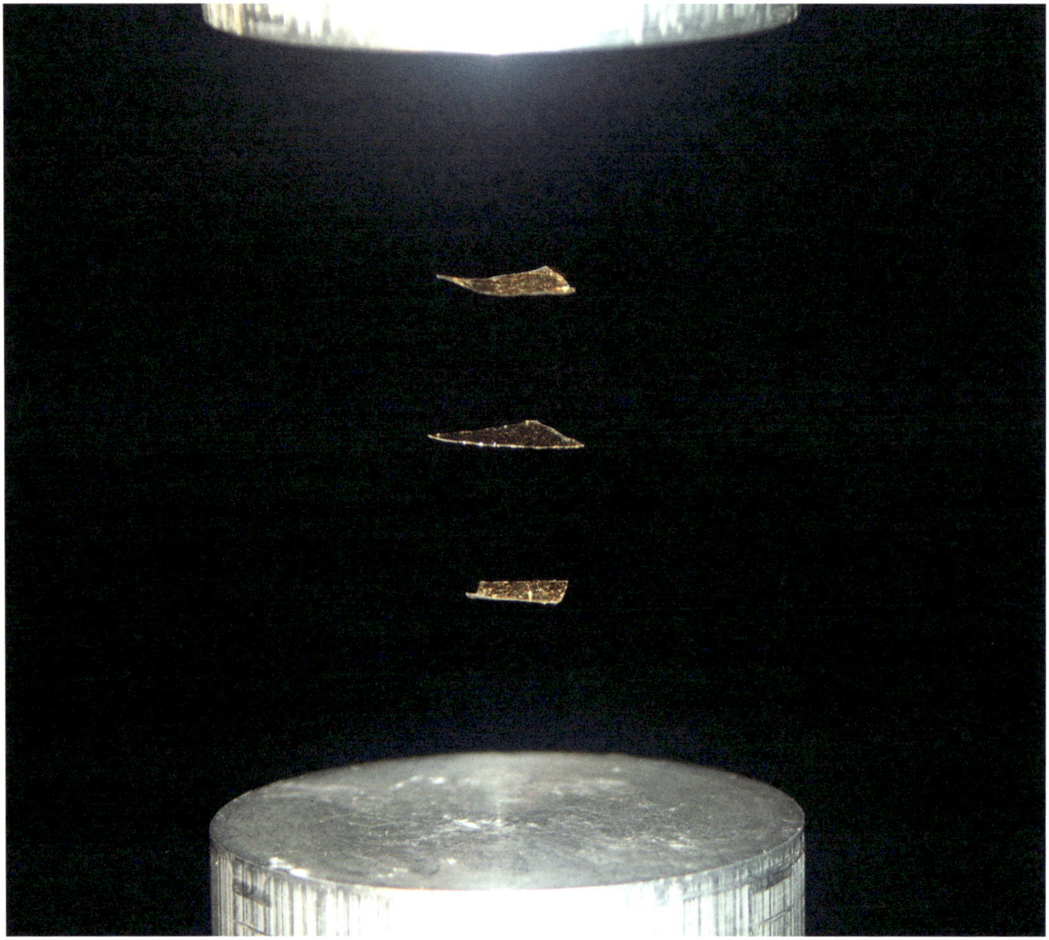

Fig. 2: Acoustically-levitated leaves of gold in Sonolevitation.

leaf's motion simultaneously affects the motion of neighboring leaves – each part is indivisible from the whole feedback system.

In *Mucilaginous Omniverse* (2009), we continue to procure weightlessness acoustically. Directly above the surface of a sonically-vibrated bath of silicone oil, droplets of the same or other less viscous liquids bounce on an acoustically-generated pillow of air. Under certain sonic conditions, droplet-lattices are created, while other conditions incite quantum-like behavior such as single particle (droplet) diffraction, tunneling and quantized orbital motion. This macroscopic analogue of quantum behavior is also imbued with an uncanny form of prebiotic memory: "the droplet moves due to its interaction with the distorted interface, this means that it is guided by a pilot wave that contains a path memory. Through this wave-mediated memory, the past as well as the environment determines the [droplet's] present motion".[1]

Photonic Wind (2013) involves light-induced levitation and migration of matter, known as photophoresis. Shining into a vacuum chamber, an Yves Klein blue laser beam levitates and propels diamond micropowder. Forming starry jets and languorous vortical clouds, the diamond dust evokes light's pervasive flow, irrepressibly transforming everything in its wake.

The artwork commenced as part of a Liquid Things residency in Vienna, at the Atom Institute and the University of Applied Arts. Fortunately, we have a physicist friend in Vienna, Tobias Nöbauer, who was our scientific consultant at the Atom Institute and graciously allowed us to work at his laboratory. Inspired by an historic photophoretic experiment (that will be described later), we initially chose to levitate silver micropowder. After further tests with diamond dust, we decided to swap the silver for diamond particles – not only because they levitated more easily, but also because the diamond powder withstood the thermal assault of the laser. The silver particles, on the contrary, would gradually melt together into clumps, creating an obstacle for long-term observations.

After the first successful experiments, our attention was caught by the orderly vorticity of the dust cloud that formed in the vacuum chamber. The tiny volume of the container might have influenced the cloud's symmetry. Tobias suggested that the spiral cloud could have been the consequence of an *optical vortex*, shaped by a toroidal aberration in the center of the laser lens. Such lenses have recently been developed for *optical tweezers* that tie laser beams into knots.[2]

Although the initial stages of experimentation were fruitful, we still needed to build a much larger transparent vacuum chamber. We chose a stitch-free, vacuum-friendly, blown glass sphere. However fond we are of the spherical shape that a body of liquid most often assumes in microgravity, the impressive ten-liter vacuum bubble, created for the first version of *Photonic Wind*, proved to be a disappointment. The numerous laser reflections traversing its curved walls prompted an entirely unintended light spectacle that completely obscured the subtleties of photophoresis.

A glass cube with six glued sides is a far inferior receptacle of emptiness. Microscopic pores in the dried glue allow air molecules to flow easily into the cube. Nevertheless, after multiple micropore tests with bubbling soap film, the cubical vacuum chamber was finally operational. Though the diamond vortex retained its relative size and quickly dissipated into a blur within the larger volume, another curious feature emerged: a diagonal dust jet appeared alongside the main particle flow, originating from the focal point of the laser beam.

Coincidentally, some of the earliest research on photophoresis was conducted in Vienna by iconoclast physicist, Felix Ehrenhaft, at the beginning of the twentieth century. His experiments with light-guided silver particles led to his conviction that he had discovered one of the missing links in our understanding of light's double nature: the phenomenon of *magnetic charge* (a hypothetical magnetic mirror of electric charge) imparted by light waves. With a microscope aimed into a vacuum chamber, Ehrenhaft observed the light of a silver vapor arc shifting the trajectories of falling silver vapor droplets. He attributed this shift to a magnetic charge, underlying photophoresis. At the same time, the unlit droplet trajectories, mobilized by an electric field, enabled Ehrenhaft to calculate the *elementary quantum of electric charge* (the charge of a single electron).[3] These constantly-refined falling droplet experiments contributed to a long-lasting controversy (1910-1923) between Ehrenhaft and Robert Millikan over the precise quantity of the elementary electrical charge. Finally, after numerous articles and debates which thoroughly stirred up the physics community, Millikan's calculation was judged to be more exact, for which he was awarded the Nobel Prize in 1923. After this, interest in Ehrenhaft's *magnetricity* quickly waned. It was a concept that dismayed his colleague Albert Einstein to such an extent that he put an end to their friendship.

In 1929, the magnetic charge hypothesis was reignited in Paul Dirac's legendary article where he posited the existence of an *anti-electron* hole with an equal mass and opposite charge to its twin. A few years later, the anti-electron, now known as the positron, was discovered in a bubble chamber by Carl Anderson.

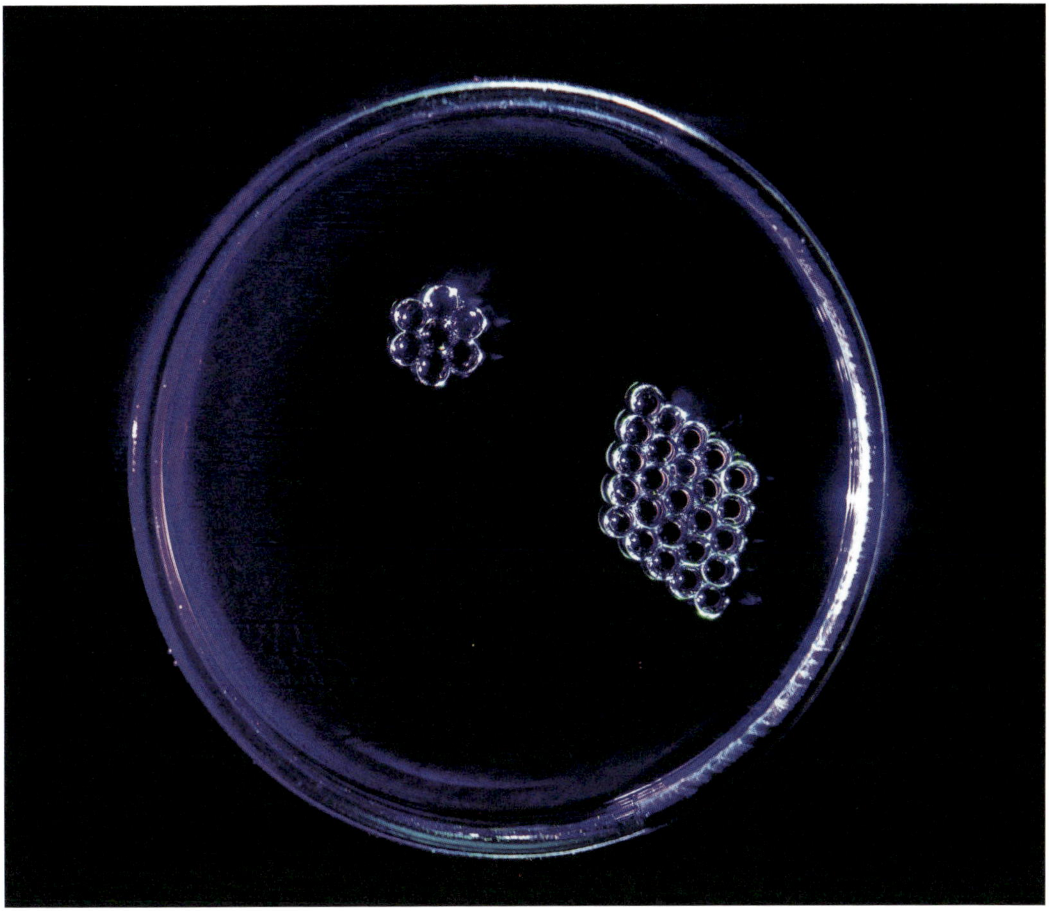

Fig. 3: Acoustic levitation of laser-lit silicon oil droplets in Mucilaginous Omniverse.

The breakthrough brought a great deal of attention to Dirac's treatise on the ephemeral nature of magnetic monopoles, which he correlated with the *nodal singularities* (alternating magnetic phases) of a quantum wave.[4] Though seemingly too ethereal to detect directly, magnetic monopoles were systematically sought among cosmic rays and meteorites as well as in particle colliders and lunar dust. The pursuit was fueled not only by Dirac's musings, but also by a growing number of grand unified theories for which magnetic monopoles were deemed quintessential.[5] Without them, it was difficult to fathom how light might

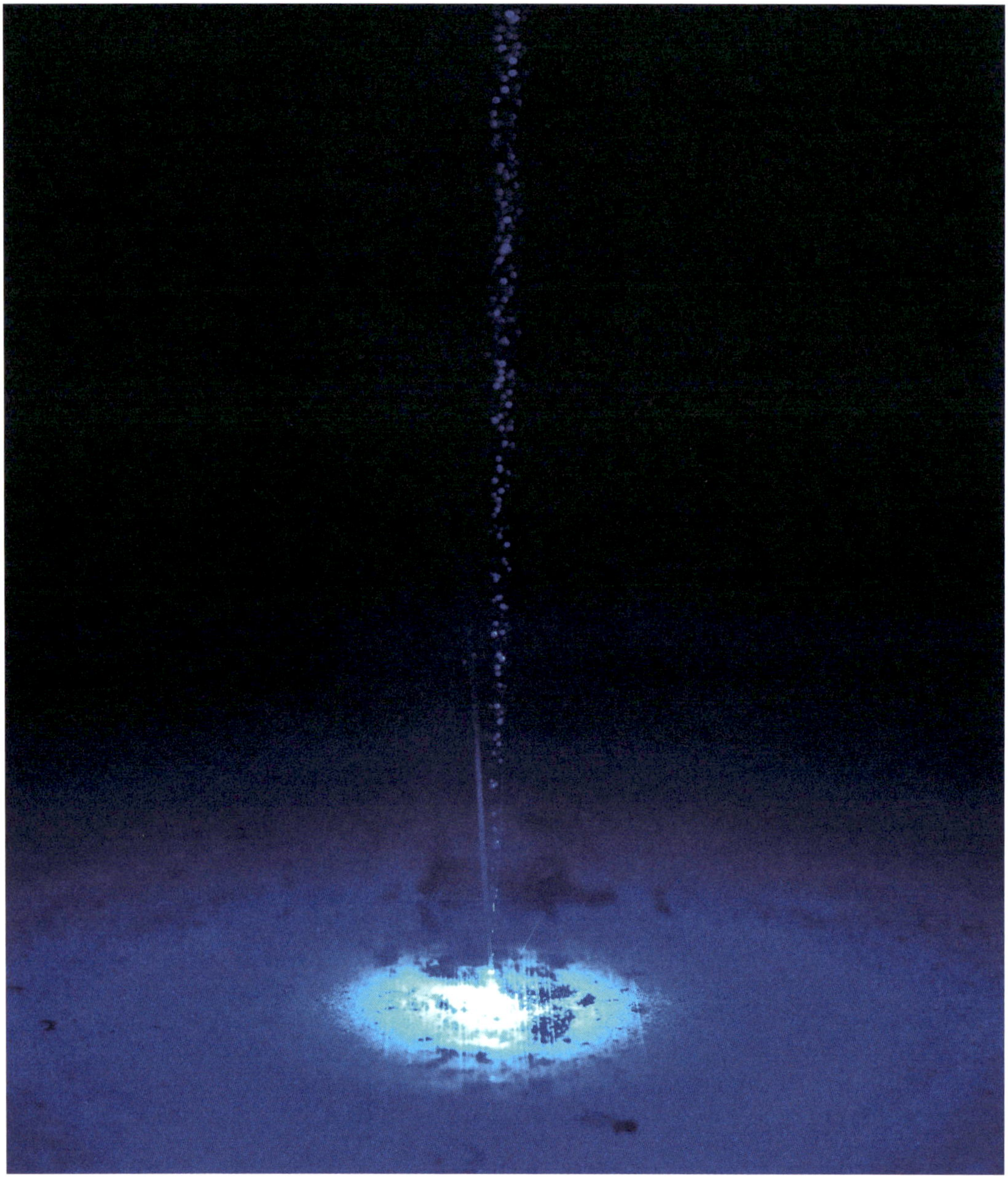

Fig. 4: Propulsive emission of diamond micropowder in Photonic Wind.

have initially decayed into electrons and positrons as it traversed the primordial cosmic vacuum. The presumably massless fluidity of light must have condensed into mass-laden *point charges* as it coupled with the dense sea of energy hiding inside the empirical void. Eventually, both the early and modern-day vacuum were proposed to be a *quantum Bose liquid* mirroring the polarized behavior of light. Alas, magnetic poles have yet to be found in the spinning vacuum's wake. However, during the last decade, magnetic monopoles have arguably been detected as momentarily *separated dipoles* in an exotic condensed metal known as *spin ice*.[6]

As the search for magnetic charge continues, photophoresis has meanwhile been suggested as the phenomenon responsible for planet formation: light-propelled rotation of cosmic dust gradually snowballs into a planetesimal, the seed of a planet.[7] Simultaneously revealing the slow-motion birth of a planet and the rapid dynamics of photo-molecular interactions, *Photonic Wind* leads us towards the very origins of fluidity – toward a multi-fingered flow into and out of the void.

1 Eddi, A., Sultan, E., Moukhtar, J., Fort, E., Rossi, M., Couder, Y.: *Information stored in Faraday waves: the origin of path memory*, in: *Journal of Fluid Mechanics*, Vol. 674, 2011, pp. 1-31
2 Padgett, M., Bowman, R.: *Tweezers with a Twist*, in: *Nature Photonics*, Vol. 5, 2011, pp. 343-348
3 Ehrenhaft, F.: *Eine Methode zur Bestimmung des elektrischen Elementarquantums, Physikalische Zeitschrift*, Vol. 10, 1909, pp. 308-310
4 Dirac, P. A. M.: *Quantized Singularities in the Electromagnetic Field*, in: Proc. Roy. Soc. A, Vol. 133, 1931, p. 60
5 Carrigan, R., Trower, W. P.: *Magnetic monopoles*, in: *Nature*, Vol. 305, 1983, pp. 673-678
6 Morris, D. J. P., Tennant, D. A., Grigera, S. A., Klemke, P. B., Castelnovo, C., Moessner, R., Czternasty, C., Meissner, M., Rule, K. C., Hoffmann, J.-U., Kiefer, K., Gerischer, S., Slobinsky, D., Perry, R. S.: *Dirac Strings and Monopoles in the Spin Ice Dy2Ti2O7*, in: *Science*, Vol. 326, Issue 5951, 2009, pp. 411-414
7 Teiser, J., Dodson-Robinson, S. E.: *Photophoresis boosts giant planet formation*, in: *Astronomy and Astrophysics*, Vol. 555 (A98), 2013

Sources of the figures:

KONTINUUM

The exhibition *Kontinuum* was shown from January 28 until February 14 2015 at the gallery *Im Ersten* in Vienna, Austria. Five artists involved in the research of *Liquid Things* presented pieces which were developed in the course of the project: Evelina Domnitch and Dmitry Gelfand (*Photonic Wind*), Aernoudt Jacobs (*Color of Noise*), Yunchul Kim (*White out*), and Roman Kirschner (*Leaking (for the time being)*).

Kontinuum aimed at a wider audience. To aid in contextualization, the concept was formulated briefly and accessibly, with the aspects of fluidity burbling under the surface:

> *The relationship between the human and non-human is currently under intense discussion. The view of the world as being subject to human exploitation and control was dominant for many centuries. But recently the idea of an abundance of natural resources in a stable material world, under human control and enduring the forces of nature, has become dated and problematic. As a consequence, matter is increasingly perceived as embedded within life processes and rich meshworks of relationships (cultural, social, chemical, physical, etc.). From these surrounding factors and exchanges it draws its properties, potentials and meanings.*
>
> *In the light of this expanded perspective on matter, the research project Liquid Things investigates creative processes that go beyond the use of finished and pre-fabricated materials. These finished materials only represent a small choice of possibilities in comparison to the material continuum that the world consists of. The continuum's full spectrum contains many more impurities, mixtures, combinations and singularities than could ever be captured by abstract philosophical notions of matter or physical/chemical categorizations. All these special cases and exceptions – which, rather, are the rule – are meeting points for encounters with the continuum's plurality and extensive complexity. The exhibition Kontinuum presents some particular processes of materialization and shows how five artists have dealt with matter's dynamics, recalcitrances and activities.*

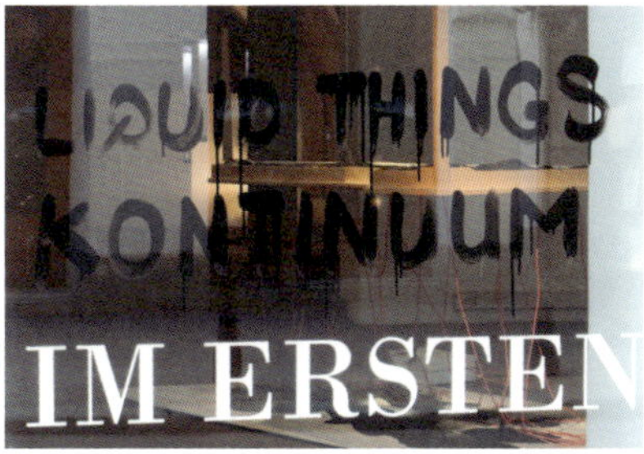

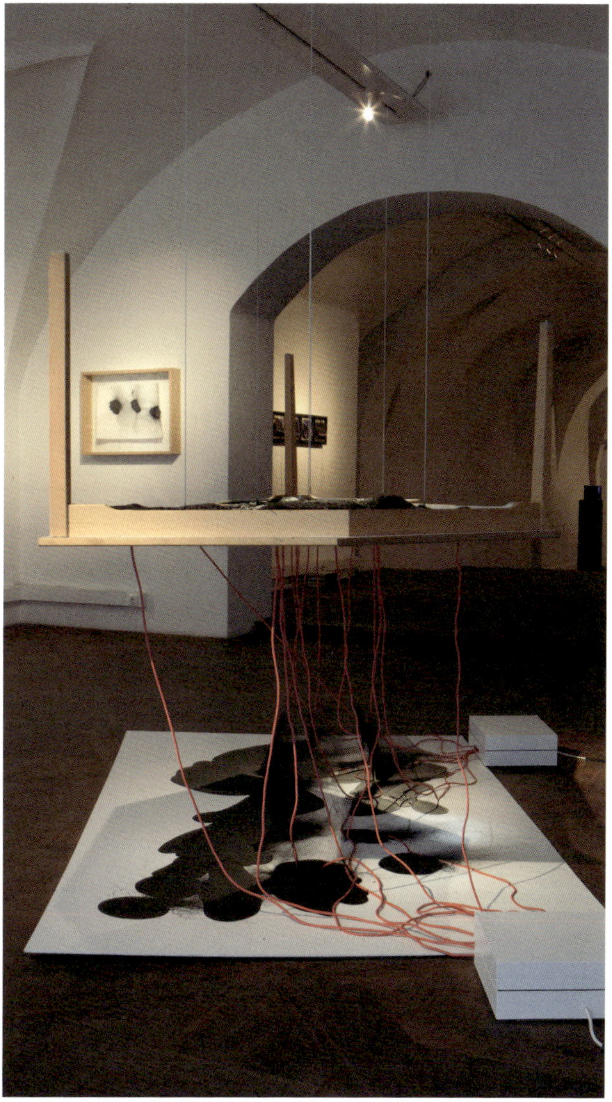

Spatial view of the exhibition Kontinuum. Foreground: *Leaking (for the time being)* (Kirschner); Second room left: *Color of Noise* (Jacobs), right: *Photonic Wind* (Domnitch & Gelfand).

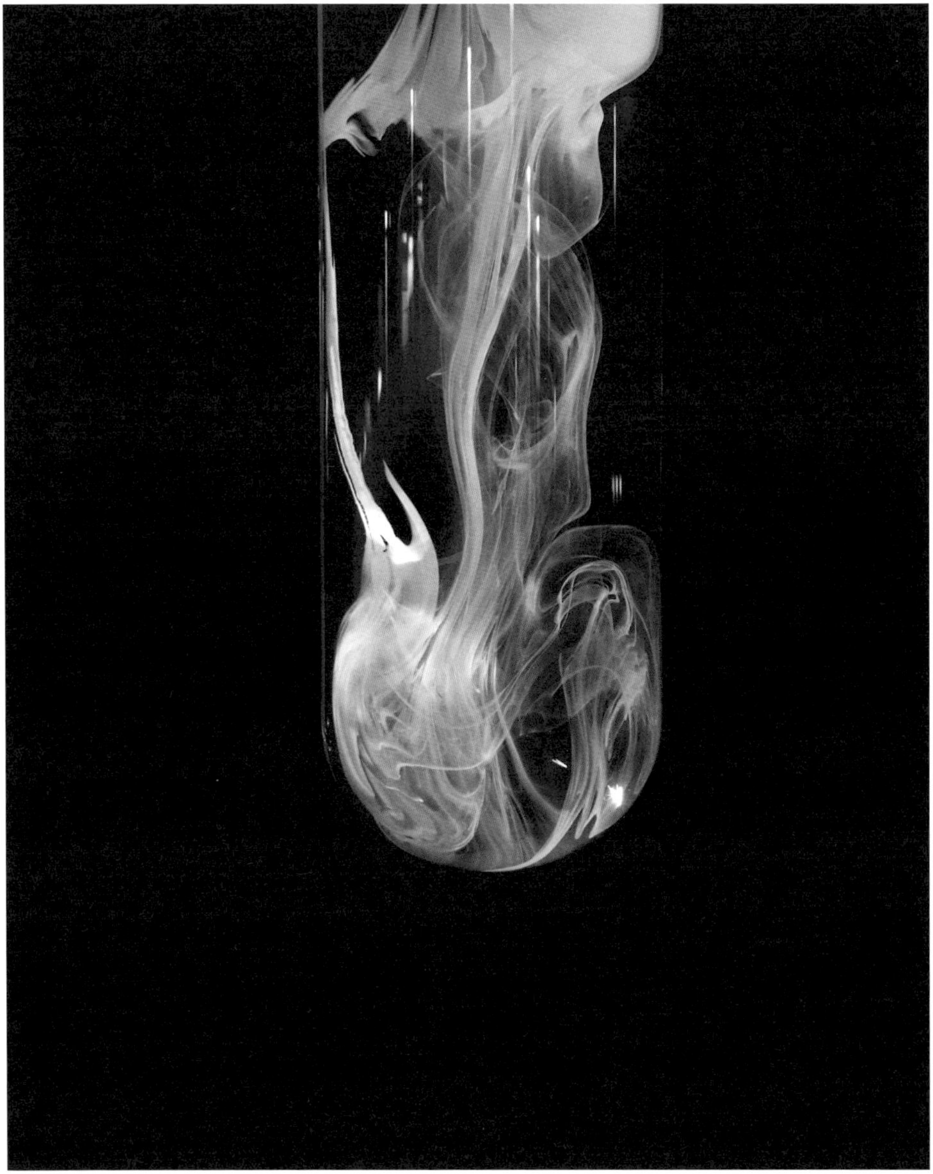

White Out (Kim) photos: Yunchul Kim

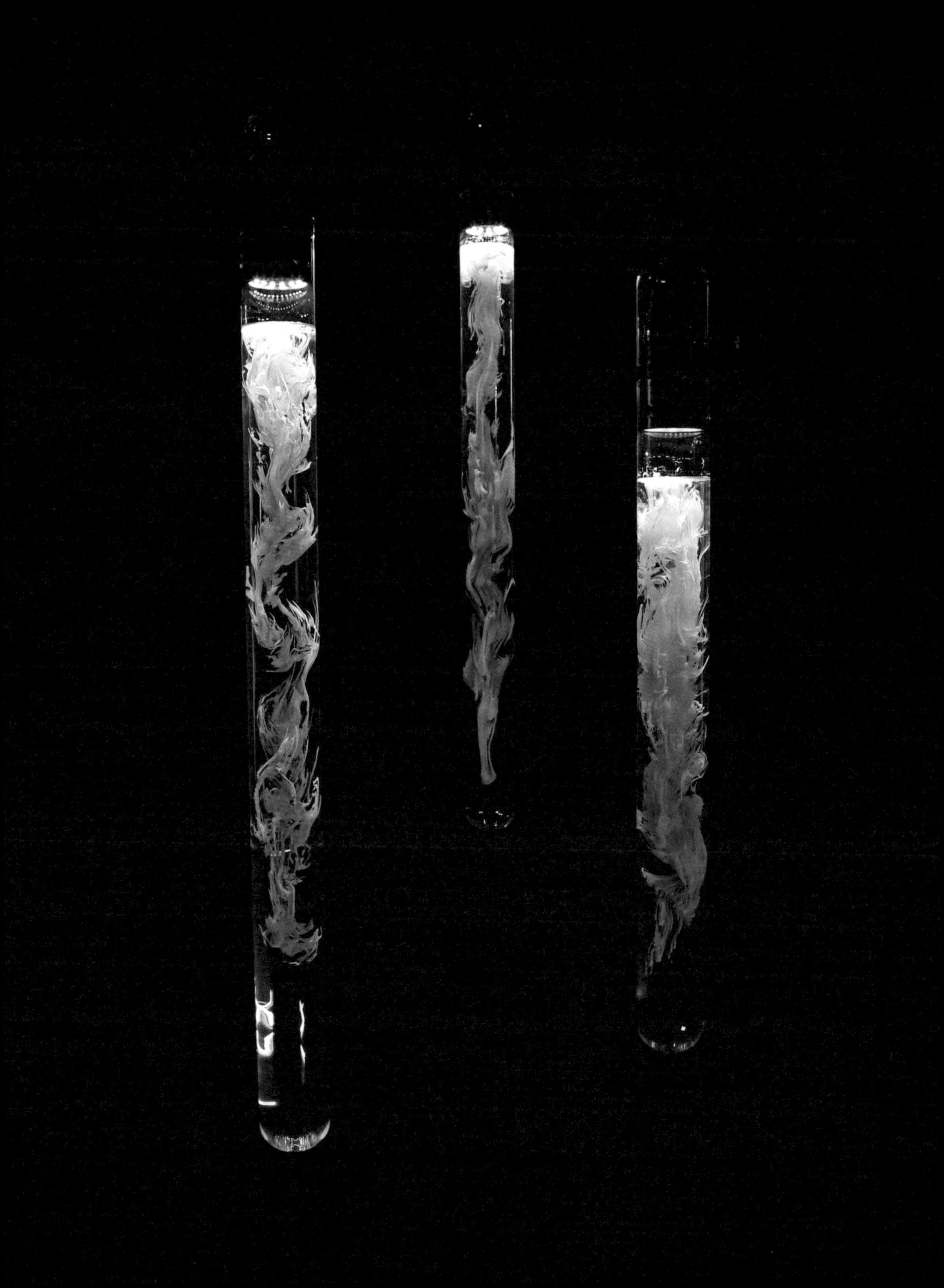

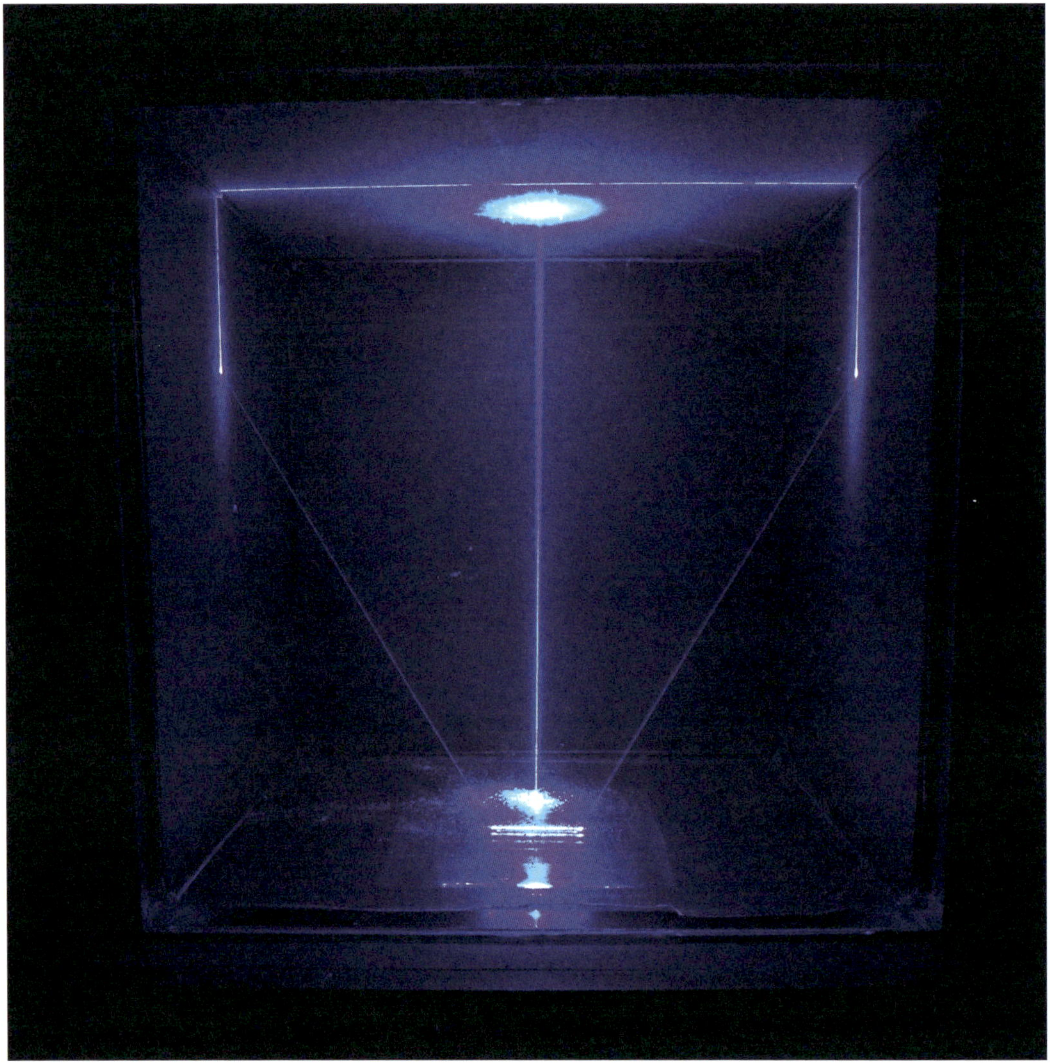

Photonic Wind (Domnitch & Gelfand)

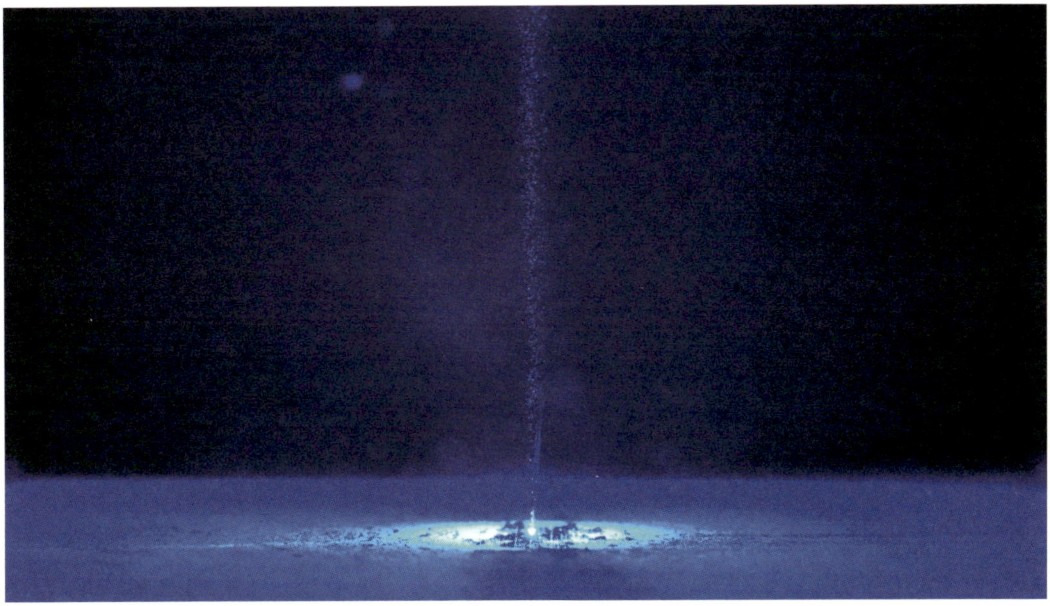

photos: Evelina Domntich and Dmitry Gelfand

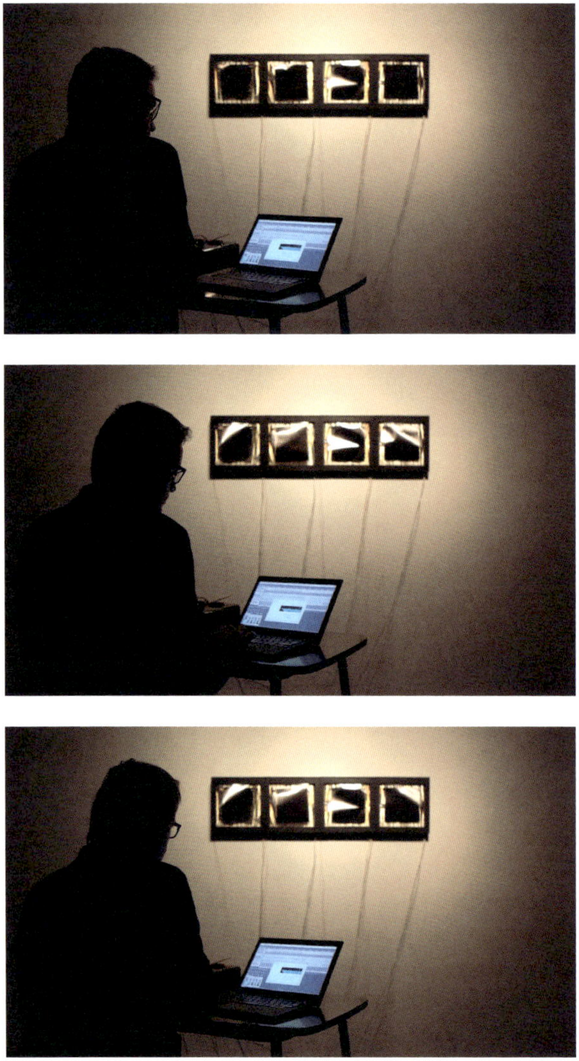

Sequence of images showing Aernoudt Jacobs during his perfor-mance of the piece Brasov using his installation *Color of Noise*.

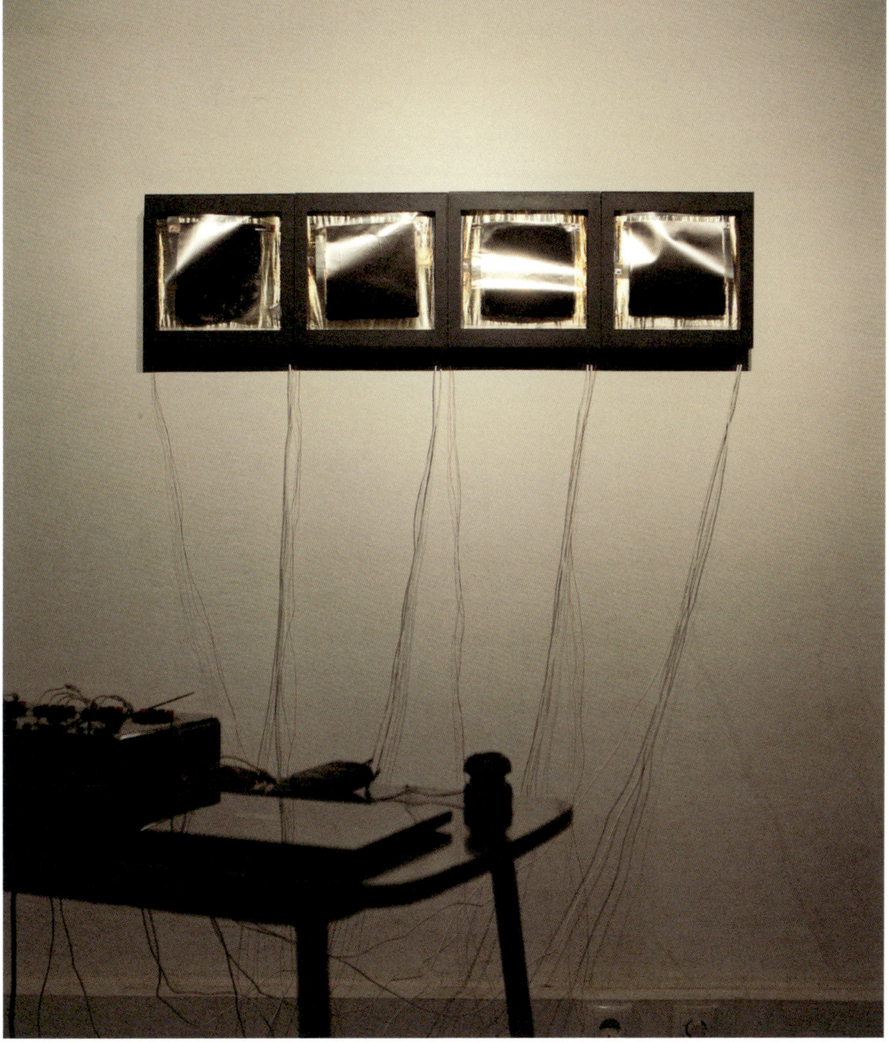

Color of Noise (Jacobs), prepared for live performance

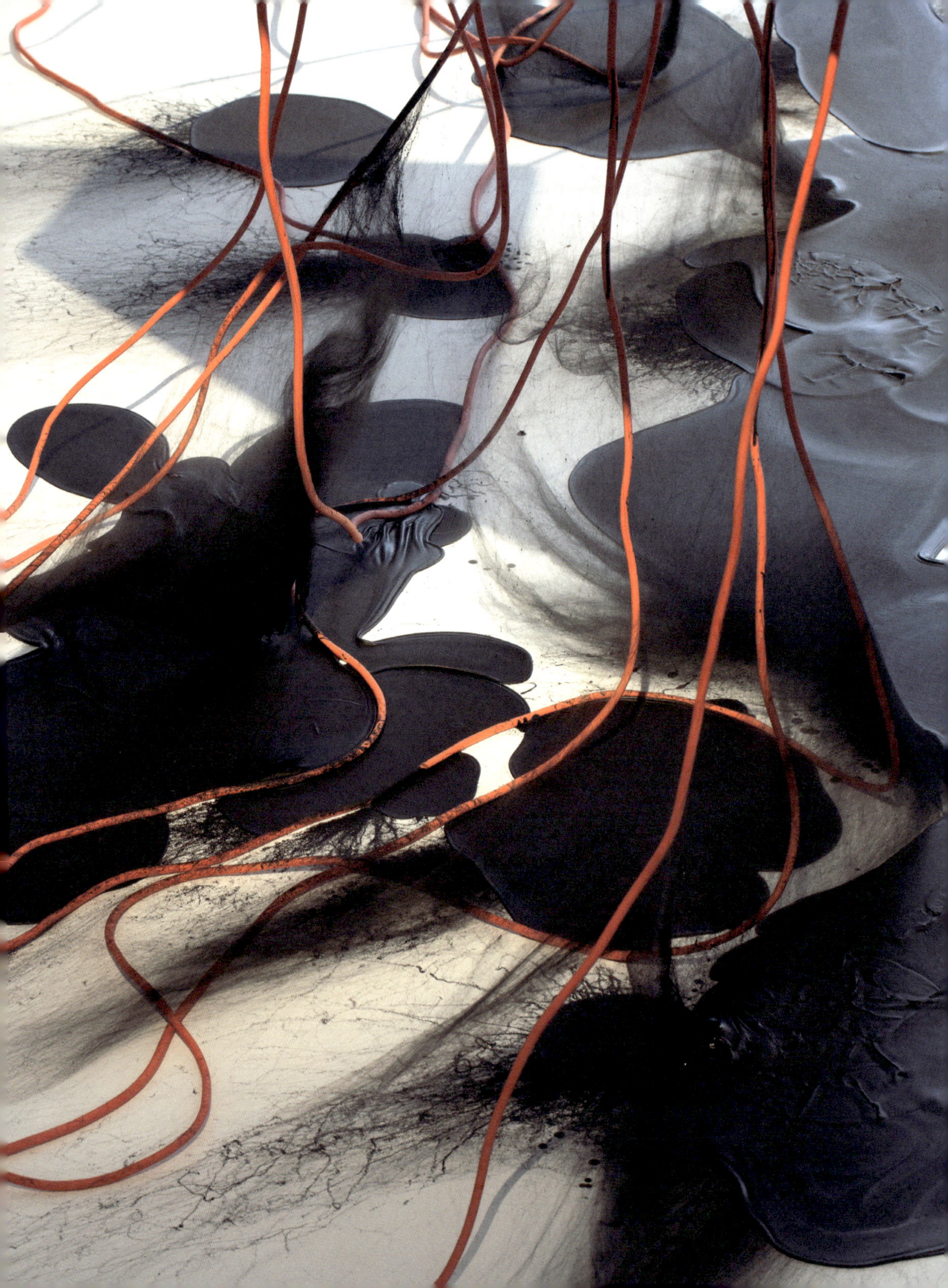

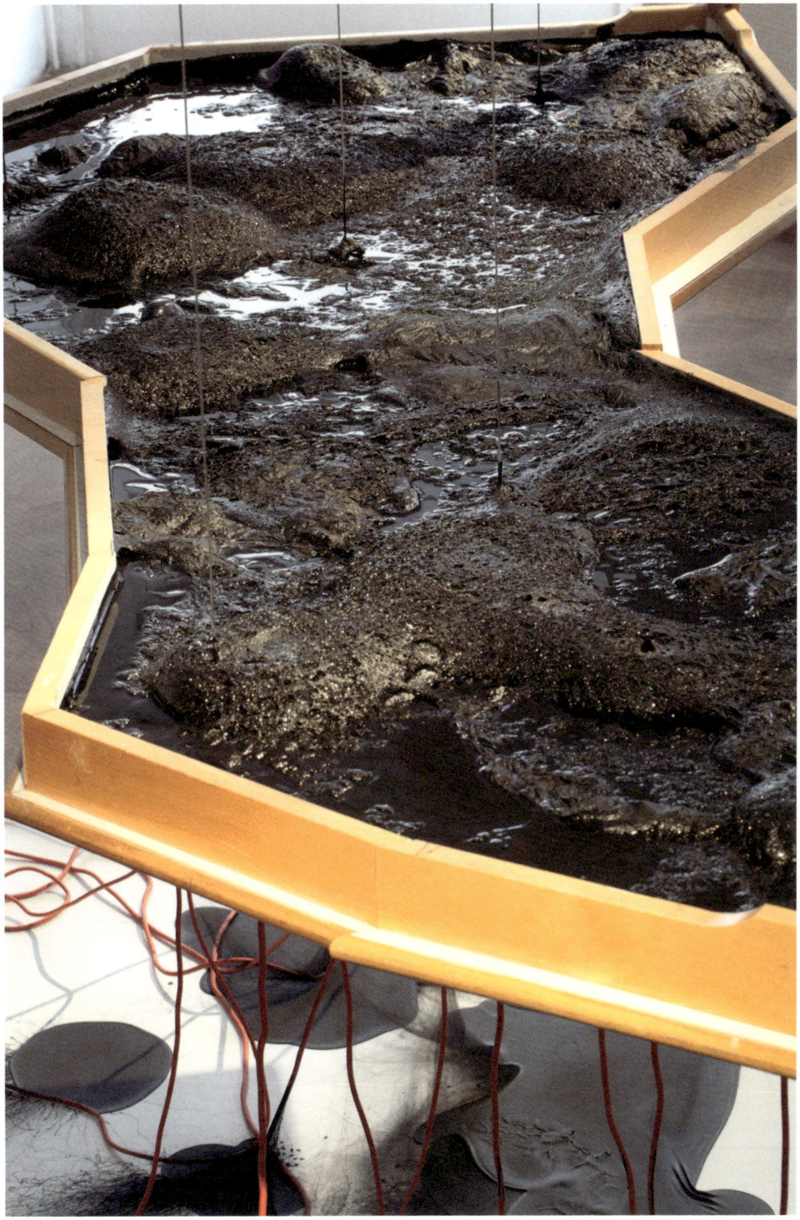

Leaking (for the time being) (Kirschner)

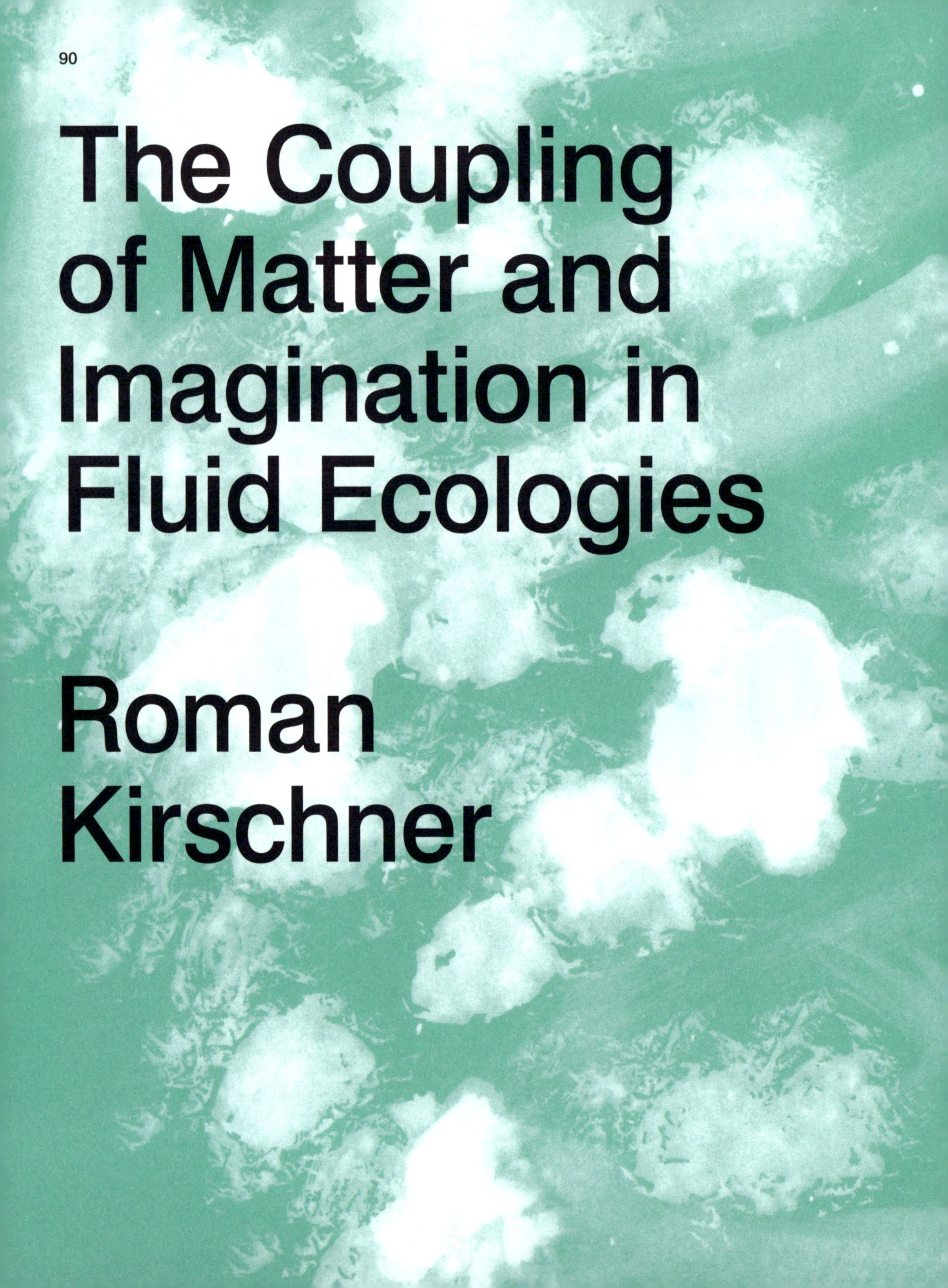

The Coupling of Matter and Imagination in Fluid Ecologies

Roman Kirschner

The world as it is is not the product of my representation; my knowledge, on the contrary, is a product of the world in the process of becoming.[1]

Michel Serres

Beginning a text with a quotation about the state of the *world* might seem pretentious, but it forces certain questions upon author and reader: Can we really make such universal statements without trepidation? Do we know and can we know enough about internal representation? Is the world really completely independent from human cognition?

The following text should be considered as nothing more than an essay that attempts to sound out the almost inaccessible yet highly complex coupling between the realm of matter and the realm of the imagination. The essay is driven by the hypothesis that there is a deep parallel between the two that may become visible when looking at the wider context and the interrelationship between environment and cognition. Accordingly, *matter* is understood as matter-energy in flux, never solid or fixed, and dependent on its environment through permeable boundaries, interrelated pathways and material flows. *Imagination* is defined as a mental faculty that can be characterized as being plastic or fluid in relation to its capacity to deform and transform visual, auditory, and motor *images* into new constellations. In this way, imagination constitutes an inextricable part of cognition, interacting with perception, memory, affect, reason, and so on. The central question to ask, in this context, relates both to the way that an individual's cognition and environment interrelate and to those principles and processes of mutual formation and information, which guide the relationship between the two. In this investigation, I am not looking for irrefutable answers; instead, I follow a presentiment that is nurtured within the discipline of arts practice, which is accustomed to dealing with the material/mental realm in an experimental and inquisitive way. The ulterior motive of this endeavor is thus not to think and write from a perspective of pure analysis or interpretation but to keep artistic production in mind. Since no explicit research into this very specific topic seems to have been conducted, the task of this essay will be to explore the area by collecting and connecting elements from related studies in fields as diverse as anthropology, philosophy, psychoanalysis, and chemistry, to name a few.

Ecological approach

In order to think the two dynamic realms of cognition and environment together and to detect parallels, intersections, and mutually catalyzing factors, it seems counterproductive to begin by describing each realm individually. Instead, it will be more helpful to start by looking at the intertwined situation of both realms from an ecological point of view. Gregory Bateson, the anthropologist and cybernetician, provides a very good entry point. He claimed that matter and mind have to be understood as being embedded in systems of energy and information exchange and not as bare and isolated entities. In this systemic exchange, mind does not emerge as something separate, but it occurs. Matter and mind are in no way separable except as *levels of description*.[2] Just as the material/energetic aspects of what he calls *units of evolutionary survival*[3] need to be understood in their milieu, the same environmental view is necessary for the mental aspect. Thus, on the one hand, Bateson wants to "include the completed pathways outside the protoplasmic aggregate, be it DNA-in-the-cell, or cell-in-the-body, or body-in-the-environment."[4] On the other hand, he states that the individual mind is not only immanent in the body; rather, it is "immanent also in pathways and messages outside of the body; and there is a larger Mind of which the individual mind is only a subsystem."[5] It is outside the scope of this essay to look at this larger Mind in more detail and to describe how the individual mind stretches out in pathways and messages. Suffice it to say that Bateson sees the mind as distributed and, more importantly, only comprehensible and viable in its environment. This establishes a very strong link between cognition and environment and constitutes the basis for any further investigation into the relationship of matter and imagination.

The importance of process and the quality of change

Before discussing the material environment and the creative imagination in more depth, it is necessary to emphasize and acknowledge the processuality and constant development of the realms of matter and mind in a way that goes further than everyday understanding. Normally, processes are thought or described as changes between distinct states. But this leaves out some very important features of these changes. For example, one knows the state of a freshly made soufflé and its state ten minutes later when it is served at the table. It has turned from an airy delight into a flat, hard tile. But what happened between start and end point? How exactly did the transformation happen? Did it collapse immediately after leaving the oven or did it slowly deflate while cooling? One might argue that it doesn't make any difference because the soufflé is inedible; however, for the evaluation of the process itself, it is very important to keep factors like the continuity of change, the timeframe, and material/energetic happenings under consideration. Questions about the quality of the change become even more obvious when looking at more complex processes, where the idiosyncrasy of a specific change is the central clue to its understanding and valorization, as is the case, for example, with a political or artistic process. Thus, it is important to keep in mind that a process is not a jump from one state to another, but a more or less soft flow in a time-space continuum. This embeddedness in space and time needs to be acknowledged explicitly. The philosopher of chemistry, Paul Needham, takes this claim even further by demanding that processes should be considered as autonomous entities that require special attention as complements to *continuants*.[6] This relates directly to the problem that in both of Bateson's *levels of description*, matter and mind, the question arises as to how constancy and becoming can be reconciled for effective description. How can things be thought in their complex processuality, in which identifiable entities are formed between materials and forces, persist temporarily in a network of flows and transformations and then inexorably fade away?

Material concerns

Needham's view is made possible by advances in the sciences from classical dynamics in an idealized and reversible (micro-)cosmos to the rather more recent theory of irreversible thermodynamics, as advocated by physical chemist, Ilya Prigogine.[7] In the nineteen-sixties and seventies he developed a theory, which describes material processes and complex systems far from equilibrium. He thus rendered accessible for scientific inquiry and exploration a world that is outside the boundaries of determinism. In the resulting world-view, order is created through fluctuations in moments close to stability or in temporary meta-stabilities. Transient phenomena like clouds, winds, flames, complex turbulences, mixtures, pulsations, and dissipative structures are considered noteworthy processes that make up nature. Henceforth, the human environment can no longer be understood as stable but turns into a world of becoming that is excitable, in constant movement, subject to chance, and irreversible. Consequently, it becomes advisable to think more in terms of tendencies or probabilities than in terms of states. This dynamic approach forms part of the basis for the contemporary lines of thought that are woven together in the *new materialisms*, which emphasize not only becoming, materialization and self-organization but which also accentuate the relationality of open systems.[8] The guiding interconnectedness of entities in their constant interdependent evolution brings about an understanding of unpredictable, non-linear developments of assemblages[9] that radically neglects any notion of causality or teleology. Their development flows can bifurcate spontaneously at any time due to more or less distant events that influence local equilibria. However, if everything is understood as being in flux and entangled in a multitude of relations, the complexity of open assemblages leads to the risk of annihilating any critical or productive potential.[10] In this situation of instantaneous, arbitrary development, it is as if everything is doomed to slip through one's fingers like mercury. The only way to get a hold on it is to try not to grab it, but to form a cup with one's hand for temporary containment and inspection. This does not, however, change the fact that it is necessary to accept oneself in the middle of things and in the middle of an unstoppable forward movement – not in the sense of progress, but in the sense of a movement toward the not-yet-known. It is in the acceptance of and concentration on this movement forward, in navigating sequences of moments that are not-yet-fixed, of things-in-becoming,[11] when connections, interlacements, and possible variations appear and disappear – ready to be chosen or neglected. The anthropologist Tim Ingold has not only explicitly acknowledged this situation, but has also started research[12] into how it is methodically possible (in his case in anthropology) to join the unfolding of the world instead of analyzing backwards from products, objects or results. He claims that the term *material agency* results from the traditional objectifying of the material realm, which ultimately turns its parts into dead objects. Thus, the material realm is comprised not of independent static objects, but of "circulations of materials that continually give rise to the forms of things even as they portend their dissolution."[13]

In further elaboration of these ideas, Ingold shifts his attention from the passive surfaces of objects to the active boundaries of things, which are only sustained thanks to the flow and exchange of matter across them. In earlier research, the biologist and neuroscientist, Francisco Varela, also dealt with this issue of boundaries, especially in relation to the development of simple cognition; he described the dialectic of living systems, using the example of bacterial cells. This dialectic consists of the organism's need to distinguish itself from the milieu that it stems from in order not to dissolve back into it. At the same time, the organism needs to maintain its coupling with the environment in order to receive nutrition and information.[14] Ingold takes this issue of permeable containment and immersion further by asserting a *leaking* of things. His claim not only makes sense with regard to entropy and dissipation but also reflects the fact that environmental interactions, and thus life processes on a material level, take place by discharge and absorption through soft and mostly active boundaries.

To sum up, when looking for the environment in which mind occurs, we are confronted with a processual material world that is guided by principles of irreversible thermodynamics rather than by transcendental ideas, causalities, and teleology. All material entities are entangled in the flows of matter and energy, through their formation processes, at the very least, as well as entropy and dissipation. Living systems are involved in the dialectics of two fundamental needs: first, to distinguish themselves from the environment using clear boundaries, and secondly, to stay connected to that same environment in order to survive. The boundaries leak in both directions, inward and outward, to support environmental interactions.

Mental concerns

According to the defender of the extended mind hypothesis, philosopher Andy Clark, the same is true for the mind in an analogical manner. He claims that the skull, or the body, is not simply a vessel for the mind. Instead, brain, body, and world collaborate closely to make the mind appear in the material flux. In his words, this sounds even more provocative. He writes that "Mind is a leaky organ, forever escaping its 'natural' confines and mingling shamelessly with body and with world."[15] Such a formulation and approach certainly provide grounds for criticism, and indeed, the extended mind theory is disputed.[16] It is not crucial for this text, however, to know exactly where human cognition is located and how far it stretches. Of greater importance is Clark's less disputed hypothesis, that human cognition – and especially real embodied intelligence – "is fundamentally a means of *engaging* with the world." [17] If this is true and cognition has evolved through interaction with the material world, while also being part of it, important questions arise as to which features, models and dynamics they may share and how.

Clark's notion of embodied cognition clearly stands against the widespread computational view of mind, in which mind is to the brain as software is to the computer on which it runs. From this perspective, mind is reduced to code that holds passive, internal representational structures and that triggers computational procedures. Clark, however, asserts that mind is not so much an issue of internal representation. Instead, he argues for active strategies that leave much of the information out in the world. He writes, "The image here is of two coupled complex systems (the agent and the environment) whose joint activity solves the problem. In such cases, it may make little sense to speak of one system's *representing* the other." [18]

A very similar anti-representational (and anti-computational) stance is taken by the proponents of dynamical systems theory. In 1995, Tim Van Gelder argued for the integration of the nervous system, body and environment, which he regarded as "complexes of continuous, simultaneous, and mutually determining change." [19]

His view of a cognitive system, therefore, relies on the fundamental embeddedness of human agents in their environment and the observation that all aspects of the wider cognitive system are undergoing continuous change. The assumed relationship between the nervous system, the body and the environment is described as *coupling*, which, in the context of dynamical systems, means that *changing only one parameter of any system component results in immediate changes to the overall dynamics*. This deep integration and direct interdependence serves as a cardinal point for any further investigation.

With these two positions, which approach the relationship of matter and mind from the mental side, we are confronted with an anti-representational stance that argues for a fundamental pairing of the mental and material realms. Cognition has been formed out of the need to deal with the material environment, to perceive it, to act in it, and to change it. However, cognition does not duplicate it fully in representation. Most information stays out in the world and both areas, world and organism, work together to resolve situations as they arise. Finally, since they are both processual and coupled, a change on one side results in a change on the other. What does this mean for imagination? What can be said, so far, is that imagination, as part of cognition and as a mental process, emerges from the interactions of body and world. Is it to be understood as non-representational, however? How or to what degree can this be achieved in a faculty that is traditionally connected to forming images? These questions may best be addressed not from the level of images but from the more abstract level of signs.

Material signs, conceptual integration, and material anchors

The neuroarcheologist Lambros Malafouris deduced from Clark's assumption about the mind's necessary engagement with the world that material culture should be "consubstantial with mind."[20] And to further back up the unity of matter and mind, Malafouris makes the effort to rethink semiotic processes. His guiding question relates to the way matter can become a sign that is meaningful for cognition. In other words, he attempts to take the argument about how signification works beyond the Saussurian view, in which the relationship between the signifier and the signified is completely independent of any material basis. He does not stop with the Peircian conception of semiosis in which icon, and even more so *index*, establish a link to the material aspects of signs and thus to the non-representational attributes of signification. Malafouris, rather, complements theories about representational signification with an *enactive*[21] approach. He postulates the expressiveness of material signs in such a way that they do not so much represent concepts as substantiate them. In this way, they rely on perceptual processes based on bodily interactions with the physical world. The concept of *weight* provides one very simple example. Its initial meaning and the entire concept can only be grasped through bodily experience, meaning that the concept of weight can only be of further use after this experience. This gives us an idea of the enactive approach without explaining the idea of material signs in its entirety. The process of involving and linking bodily experience with the semiotic conflation of cognitive and material levels constitutes the enactive approach to signification. At its core, some cognitive assimilation takes place. I now turn to the question of how this might work.

In his own concise words, Malafouris describes the enactive approach to signification as "a process of embodied, conceptual integration responsible for the co-substantial symbiosis and simultaneous emergence of the signifier and the signified that brings forth the material sign."[22] This is not easy to digest. He speaks of an embodied and conceptual integration. Again, the body plays a role in acquiring and producing information. It is not an abstract or purely mental process. The *co-substantial symbiosis* suggests that all entities, perceiver and perceived, exist in the realm of substances through which is communicated more than the necessarily extracted difference, which constitutes the sign. This background information is present while and before the sign emerges from the undifferentiated pool of substances as a meaningful something for the perceiving body.

Material signs – understood as artifacts or objects in Malafouris' neuroarcheological context – communicate through many channels to which Malafouris does not pay special attention. However, their physical, chemical, and dynamical properties are of central importance. They can react, decay, withstand environmental influences, show recalcitrance, and incorporate qualities that affect sensible organisms on non-discursive levels. In short, through their properties and behavior they can induce meaning and thus have a direct influence on signification. In addition, cultural meanings are attributed to these signs (to their materiality or to the experiences necessary to access them), together with knowledge about their coming into being, a multitude of bodily experiences that are formed earlier (and might require renewal or actualization), and much more. However, Malafouris is able to identify the core mechanism, which links all these different layers of material signs to their joint embodied experience, namely, *conceptual integration.*

Conceptual integration was described in the works of Fauconnier and Turner[23] as an explanation of how conceptual spaces can be blended together into a new combined space. In this way, the input spaces provide structure to the blend, which itself develops an emergent structure that is loosely based on the combined inputs.[24] Although the principle of conceptual integration was initially developed as an internal cognitive operation, it was later extended by Edwin Hutchins to include *material anchors,* also referred to and used by Malafouris. In Hutchins' extension, one of the conceptual input spaces is directly rooted in a material setting that can feed its own material/environmental structure and dynamics into the blend.[25] Hutchins' theory of material anchoring is very convincing and is backed up by many examples. His illustrations include blends with varying ratios of mental and material resources. They extend from the construction of fictive motion in literature to Japanese hand calendars and the complex Micronesian navigation between islands, in which the canoe is imagined to stand still while the destination island travels. In principle, the way that new meaning is created happens in the same way that Fauconnier and Turner describe the composition of input spaces based on already available knowledge of "background frames, (and) cognitive and cultural models."[26] However, they are also connected to or projected onto material structures and entities. The resulting output space is such a deep amalgamation of different conceptual and material components that *representation alone does not suffice to describe the created relationship between signifier and signified*. Hutchins further explains that humans are able to imagine material structures after becoming familiar with them. Their imaginative abilities do not stop there but extend eventually to imagining systematic transformations to imagined material structures. Finally, Hutchins claims that rationality as a cultural construct may depend heavily on the ability to create materially anchored blends. In the context of this text, a curious indicator can be found in the term anchor, which points to the intention of temporary fixation in a stream of change. Hutchins uses the expression *sufficiently immutable* to describe material structures that can stabilize conceptual elements while other mental operations are executed[27]. This fact, in turn, exposes the asserted parallels between the material and the mental realm, namely the malleability, mutability, and becoming of both material signs and cognition.

I will now summarize this rather complex section. After questioning the non-representational aspects of imagination, I have described an approach that complements the existing theories of representational signification with enactive and embodied components. In this approach, both experience through action and the specific substantiation of concepts in material signs are relevant for semiosis. Extra emphasis is placed on material properties, behaviors and histories. After looking at how cognitive assimilation may work with this understanding of signification, I introduced the theory of conceptual integration as a blend of different conceptual spaces. As an extension to that, I then introduced the idea of material anchors that can be integrated into conceptual blending networks. Together, they are a good example of a deep collaboration between and amalgamation of conceptual and material components in cognitive processes beyond representation. They play a significant part in contemporary theories of imagination, as I show in the next section.

Cognitive fluidity and double-scope blending: toward material imagination

To show the scope and the influences of the aforementioned theory of conceptual blending, I start by describing *cognitive fluidity*, a universal cognitive feature in humans that supposedly emerged from a complex material culture. Stephen Mithen coined this term during his investigations into the prehistory and evolution of mind.[28] It describes the human mind's ability to combine independent domains of knowledge, initially developed to solve very specific problems, to deal with new situations. The development of this ability seems equally observable on a phylogenetic and ontogenetic level. In ontogenesis, to give a brief example, children quickly develop distinct domains of intuitive knowledge in close interaction with the material and social environment – supposedly physics, biology, psychology, and language. Later, these domains can and will work together with behavioral and other domains. The argument behind cognitive fluidity is that experiences made in one domain need to be able to flow into others, where they can be used to find new solutions and behaviors. Cognitive fluidity allegedly represents an important precondition for mental activities, such as imagination, which requires the simultaneous use of different cognitive domains. The concept of *bisociation* resonates in Mithen's account of cognitive fluidity. The writer Arthur Koestler developed bisociation as a central mechanism in creation. The term bisociation, which is chosen in contrast to *association*, describes mental activities that connect different planes of thought or *matrices*, which can be any set of rules.[29] In Koestler's sense, to bisociate means to integrate ideas, memories, representations, stimuli, and the like, with incoherent or unconnected matrices in order to form new constellations or forms. Unfortunately for the intentions of this text, Koestler failed to mention any connections to material or bodily influences. Instead, bisociation, which is still considered an important figure of thought in creativity research, has served as the source for conceptual blending by Fauconnier and Turner. In fact, the two cognitive scientists refer to the most powerful blending network as *double-scope blending*, which clearly shows the influences of Koestler's bisociation.

Double-scope blending, which Turner calls the *engine of the human imagination*,[30] draws its creative power from being able to resolve tricky clashes between highly heterogeneous inputs. The solutions are based on an emergent logic that is able to merge, contain, and overcome differences in content and topology. Together with the other main types of blending networks, namely *simplex*, *mirror*, and *singlescope*, they represent prototypes along a continuum on which traditionally incommensurable varieties of meaning, such as categorizations, analogies, counterfactuals, metaphors, and rituals are situated.

So, if double-scope blending is able to drive imagination, can draw its scaffolding from a writer's theory of imagination, and is even the key to integrating material anchors, it certainly represents a very powerful mechanism with which a non-romantic material imagination can be built. If one then thinks beyond material anchors as a way to match concepts and structures, focusing instead on simpler but still very complex material experiences, a contemporary account of material imagination seems even easier to achieve. These material experiences – on a more substantial and less structural level, as outlined above in connection with material signs – help integrate the plasticity, dynamics and singularities of the material world into cognition and thus into imagination. Acknowledging the fact that conceptual integration networks can operate on highly heterogeneous inputs expands the resulting notion of imagination beyond the visual and also facilitates the integration of auditory or motor content, and the like. Furthermore, as well as integrating material and mental worlds, it also facilitates their coupled evolution and becoming. This is possible because the integration of two or more however different input spaces can lead to emergent results – especially when facing situations of seemingly unresolvable combinations, morphs, or juxtapositions. This fundamental and combinatorial openness and creative emergence is able to account for the productive capacity that a theory of material imagination requires.

Any reconsideration of the entanglements of matter and imagination should not, of course, omit the only explicit theory of material imagination to date, developed by philosopher Gaston Bachelard. Although less systematic than passionate and allusive, Bachelard wrote several books about how the historical four elements of fire, water, air, and earth have shaped literary images. In general, this seems a reasonable endeavor taking into account that human self-understanding was historically formed by experience and immersed in these elements. These, in turn, served as media for the projection and expression of feelings, passions, hopes, and fears.[31] Unfortunately, however, the project tended too much toward neo-romanticism. Nevertheless, after the completion of his books, which treated each of the four elements, Bachelard went so far as to compare the poetic imagination with the flame of a candle, briefly described as "a becoming-being, a being-becoming,"[32] in which matter flows visibly into new forms. In specifying that they can both be snuffed out by a breath of wind and relit by a tiny spark, he acknowledges their fragility. Additionally, they are connected through a continuous process of metamorphosis. This comparison is thematically interesting because the flame of a candle is a very illustrative example of a dissipative structure and an open system through which energy flows.[33] It is kept in a state far from thermodynamical equilibrium in which heat, light, and other substances are emitted, while its extinction is prevented by the oxidation of new wax provided through the wick. The flame of a candle, and thus the mesmerizing phenomenon to which Bachelard's poetic attention is attracted, is rendered possible by a subtle network of energetic pathways, chemical transformations and phase transitions in multilateral interdependence. This thermodynamical level of observation sheds an extra light on imagination, which Bachelard might not have considered or chose to omit,[34] namely, the perspective of seeing material imagination in a larger framework of matter/energy flows in which it helps influence and form the material environment. In this light, and in keeping with the notion of the environmental embeddedness of cognition and imagination, it seems worthwhile to reread Bachelard's oeuvre and reevaluate some of his key concepts. For example, in order to delve more deeply into the concept of reciprocity between world and imagination, we might rethink his words, "when the imagination speaks, which speaks, I or the world?"[35]

In this section, I have suggested an approach toward a theory of material imagination that builds on the idea of cognitive fluidity as a precondition for imagination, and which draws from Koestler's theory of bisociation as a method of connecting different planes of thought. I introduced double-scope blending as a form of cognitive integration network. This can be a versatile basis for developing a theory of material imagination in combination with a less structural and more substantial understanding of material anchoring that integrates the plasticity, dynamics, and singularities of the material world. Referring to Bachelard's material imagination, I proposed looking at material imagination in a larger framework of matter/energy flows in which it helps influence and form the material environment. The final challenge for this essay is to tackle the question of how the collected insights relate to arts and research and the horizon of possibilities and actions that this implies.

Material imagination in art-based research and the arts

In contrast to Bachelard's neo-romanticism, the science historian Hans-Jörg Rheinberger approaches the topic of the interrelationship of matter and imagination from a more rational angle. In dialogue with the editor of an anthology discussing the applicability and relevance of his notion of experimental systems for art-based research, Rheinberger describes experimental systems as "exteriorized spaces of imagination,"[36] in which the exteriorization facilitates the workability and interactivity of the experimental system in focus. He also discusses the value of an experimental spirit that goes beyond being a mere mental activity by placing the experimenter's interaction with the material at its core. He writes that

> *If one is not immersed in, even overwhelmed by, the material, there is no creative experimentation. In the course of the interaction with the material with which one works in an experiment, the material itself somehow comes alive. It develops an agency that turns the interaction into a veritable two-way exchange. It's both a forming process and a process of being informed.*[37]

This means that in each experiment, the structures, properties and processes of the material systems at stake influence the way that experimenters think, imagine, and eventually manage to extend the range of possibilities. It also suggests that the very specific way the world is being disclosed actively shapes the type of thoughts that are induced, and participates in determining which goals are pursued and how.[38] Although this summary might sound almost trivial in its necessary brevity, it nevertheless contains an important shift toward a nonhuman perspective[39] with regard to the part materials play. It also emphasizes that meaning is not entirely mental but also strongly connected to the outside world. In my opinion, Rheinberger's thesis that the interaction of an experimenter with its materials is a "forming process and a process of being informed" is crucial for conceiving a contemporary attitude for artists toward their materials. This two-way interaction provides sufficient space for the world's singularities and dynamics to imbue the imagination, for the materials to be reorganized in ever new constellations by the imagination (from the atomic to the societal level), and, subsequently, for the entanglement of matter and imagination to evolve.

But are the singularities and dynamics of the material world used in the arts in a challenging, experimental, and self-reflective way? The answer is that they are, to a certain extent, but there is always space for more. This is especially so with regard to situations in which the dynamic aspects of materials or substances and, specifically, their transformative capacities are adequately and processually integrated. In recent years, a number of artworks featuring such active material systems and experiences have appeared, as, for example, in Nina Canell's *Evaporation Test Series*. More frequently, however, materials are shown in passive roles staging their conceptual connectivity, for example, in Pamela Rosenkranz's installation *Our Product*. Material metaphors and references are conspicuously abundant, especially those with fluid and processual characteristics, such as Hito Steyerl's movie *Liquidity Inc.,* or in the works of the youngest generation of artists associated with post-internet art, such as Joey Holder, Adham Faramawy, Timur Si-Qin, or Kari Altmann.[40] The latter are natives of the digital world, which on the one hand provides many abstracted – so called *dematerialized* – experiences, but on the other offers services whose names suggest fluid phenomena like streams, torrents, and clouds to support instantaneous exchange and a non-hierarchical interlinkage of heterogeneous information. In this largely dematerialized sphere full of fluid metaphors, new thought patterns and modes of collaboration emerge. As Kari Altmann described from within the post-internet art scene,[41] the increasing dynamics of reposting images, visual connections and undifferentiated content sources leads to an ambiguous level of unreflectedness and turns the contributors of this scene into predictable algorithms. On the positive side, this collective, unreflected effort of instantaneous reposting leads to a proliferous complexity and is what one may call *post-conceptual*. This, in turn, is a very interesting development considering the strife for the *pre-conceptual, pre-symbolic*, or *pre-analytic*, respectively, which dominated the raw material experiences of the visual arts until the late nineties.[42]

Nevertheless, from the post-conceptual, semi-automatic, and collective excess of recombinations and amalgamations of content[43] in an environment with a reduced offer of material experiences, an urgent question arises as to which materializations (materials, material signs/artifacts, and further material experiences) may be generated to serve as compatible and challenging material counterparts for minds whose thought patterns evolved in advanced information technologies? The avalanche of 3D-printed sculptures that Jörg Heiser, editor-in-chief of Frieze magazine, has bemoaned in connection with the gallery output of post-internet artists[44] does not appear to provide an adequate answer. However, as the phenomenon of post-internet art seems to be rather short-lived, the question remains relevant and should even be extended. My concluding question thus goes beyond materializations for minds that have been amplified by information technology and telecommunications and asks how the coupled evolution and becoming of material and mental worlds can be thought and developed forward and how these thoughts, their respective materializations, and their integrated dynamics and flows can communicate through the boundaries of art institutions and their ambivalent value-generating and rigidifying formats.

1 Serres, Michel, *Hermès IV: La Distribution, Minuit*, 1977, pp. 158-157, as cited and translated in Prigogine, Ilya and Stengers, Isabelle, *Postface: Dynamics from Leibniz to Lucretius*, in: Serres, M., Hermes. *Literature, Science, Philosophy*, Johns Hopkins University Press, 1982.
2 Bateson Gregory and Bateson, Mary Catherine, *Angels fear. Towards an epistemology of the sacred*, MacMillan 1987, p. 18.
3 The unit of evolutionary survival is not the organism or the species as in Darwin's evolutionary theory but "a flexible organism-in-its-environment." If this also includes the interactions between organism and environment, *"the unit of evolutionary survival turns out to be identical with the unit of mind,"* in: Bateson, Gregory, *Steps to an ecology of mind*, Chandler Publishing Company, 1972, pp. 458, 489.
4 Ibid., p. 466.
5 Ibid., p.468; Bateson describes the larger Mind as being comparable to God but still immanent in the total interconnected social system and planetary ecology.
6 Needham, Paul, Process and Change: *From a Thermodynamic Perspective*, in: *British Journal for the Philosophy of Science*, No. 64, Oxford University Press 2013, pp.395-422; *continuants* (individuals or quantities of matter) are what remains more or less the same in change.
7 See e.g., Prigogine, Ilya, *From being to becoming: time and complexity in physical sciences*, Freeman 1980.
8 See e.g., DeLanda, Manuel, *Nonorganic Life,* in: Crary, Johnathan, et al. (eds.), *Incorporations (Zone 6)*, Zone 1992; see also Bennet, Jane, *Systems and Things. On Vital Materialism and Object-Oriented Philosophy*, in: Grusin, Richard (ed.), *The nonhuman turn*, University of Minnesota Press 2015.
9 The term *assemblage* is used here for constellations "of singularities and traits deducted from the flow"; quotation from Deleuze, Gilles and Guattari, Félix, *A Thousand Plateaus. Capitalism and Schizophrenia*, University of Minnesota Press 1987, p. 406.
10 See Coole, Diana, *New Materialism. The Ontology and Politics of Materialisation*, in: Stakemeier, Kerstin and Witzgall, Suzanne (eds.), *Power of Material – Politics of Materiality*, Diaphanes 2014.
11 See e.g., Massumi, Brian, *Semblance and Event. Activist Philosophy and the Occurent Arts*, MIT Press 2011, p. 1.
12 *Knowing From the Inside: Anthropology, Art, Architecture and Design*, (ERC Advanced Grant 2013-2018). Retrieved Oct. 9, 2015, from http://www.abdn.ac.uk/research/kfi/
13 See also Ingold, Tim, *Bringing Things to Life: Creative Entanglements in a World of Materials* (2008), p.7. Online version from 2010, retrieved Aug. 12, 2015, from http://eprints.ncrm.ac.uk/1306/1/0510_creative_entanglements.pdf
14 Varela, Francisco, *Autopoiesis and a Biology of Intentionality,* in: McMullin, Barry (ed.), *Proceedings of the Workshop Autopoiesis and Perception*, held in Dublin 1992. Retrieved Oct. 9, 2015, from ftp://ftp.eeng.dcu.ie/pub/alife/bmcm9401/varela.pdf
15 Clark, Andy, *Being There. Putting Brain, Body, and World Together Again*, Bradford Books 1998, p. 53.
16 See e.g., Adams, Fred and Aizawa, Ken, *The Bounds of Cognition*, Blackwell Publishing 2010.
17 Clark 1998, p. 98.
18 Ibid., p. 98.
19 Van Gelder, Tim, *What Might Cognition Be, If Not Computation?,* in: *Journal of Philosophy*, Vol. 92, No. 7 (July 1995), p. 373.
20 Malafouris, Lambros, *The Cognitive Basis of Material Engagement: Where Brain, Body and Culture Conflate,* in: DeMarrais, Elizabeth, Gosden, Chris and Renfrew, Colin (eds.), *Rethinking materiality. The engagement of mind with the material world,* McDonald Institute for Archaeological Research, University of Cambridge 2004, p. 58.
21 *Enaction* is a paradigm within the cognitive sciences overlapping with embodied and situated cognition, which proposes that experience is created through action. See e.g., *Defining the Enactive Approach,* in: Varela, Francisco, et al., The embodied mind. Cognitive science and human experience, MIT Press 1991, pp. 205-207.

22 Malafouris, Lambros, *How Things Shape the Mind*, MIT Press 2013, p. 99.
23 Fauconnier, Gilles and Turner, Mark, *The way we think: Conceptual blending and the mind's hidden complexities*, Basic Books 2013.
24 Fauconnier, Gilles and Turner, Mark, *Conceptual Blending, Form and Meaning*. Retrieved Oct. 20, 2015, from https://www.researchgate.net/publication/45359086_Conceptual_Blending_Form_and_Meaning
25 Hutchins, Edwin, *Material anchors for conceptual blends*, in: *Journal of Pragmatics*, Vol. 37, 2005: pp. 1555-1577. Retrieved Oct. 20, 2015, from http://hci.ucsd.edu/hutchins/documents/MaterialAnchorsPragmatics.pdf
26 Ibid., p. 1556.
27 For an in-depth description of a dynamic example in action, see, e.g. people queuing for tickets, ibid., p.1559-62.
28 Mitten, Stephen, T*he prehistory of the mind: a search for the origins of art, religion, and science*, Thames and Hudson 1996.
29 Koestler, Arthur, *The Act of Creation*, Picador 1975.
30 Turner, Mark, *The Way We Imagine,* in: Roth, Ilona (ed.), *Imaginative Minds*, British Academy & Oxford University Press 2007.
31 See also Böhme, Gernot und Böhme, Hartmut, *Feuer, Wasser, Erde, Luft–Eine Kulturgeschichte der Elemente*, CH Beck 2010.
32 Bachelard, Gaston, *The flame of a candle*, (trans. Joni Caldwell), Dallas Institute for Humanities & Culture 1988, p. 24.
33 An updated, contemporary comparison would show parallels between double-scope blending and *neuroplasticity*, the brain's ability to change in both physical structure and functional organization in response to neural activity, emotions and experience (including changes in environment and behavior).
34 Bachelard kept his epistemological works strictly separate from his writings on material poetology, which initially emerged from an attempt to reveal pre-scientific mindsets.
35 Bachelard, Gaston, *La poétique de la rêverie*, Presses Universitaires de France, 1960 p. 161.
36 Rheinberger, Hans-Jörg, *Forming and Being Informe*d, in: Schwab, Michael (ed.), *Experimental Systems: Future Knowledge in Artistic Research*, Leuven University Press 2013, p.210.
37 ibid., p.198
38 This seems equally valid for the material target layer and the technical support layer of experimentation. See also Aydin, Ciano, *The artifactual mind: overcoming the 'in-side-outside' dualism in the extended mind thesis and recognizing the technological dimension of cognition,* in: *Phenomenology and the Cognitive Sciences*, Springer 2015, Vol.14, pp. 73-94.
39 See also e.g., Grusin, Richard (ed.), *The Nonhuman Turn*, University of Minnesota Press 2015.
40 See Tabbush, James, *Like Bruce Lee Says: Be Water*, in: *Elephant Magazine*, Vol. 23, Summer 2015.
41 Burke, Harry and Altmann, Kari, *Artist Profile: Kari Altmann*, interview for *Rhizome*, March 2015. Retrieved Nov. 2, 2015, from http://rhizome.org/editorial/2015/mar/25/artist-profile-kari-altmann/
42 See Schawelka, Karl, *'More matter with less art?' Zur Wahrnehmung von Material*, in Wagner, et al. (eds.), *Material in Kunst und Alltag*, Akademie Verlag 2002.
43 This frenetic, post-conceptual reposting of heterogeneous content can be seen as an example of externalized cognitive fluidity. Analogous to the role of cognitive fluidity as a precondition for the formation of a recombinatorial and productive imagination, this kind of externalized cognitive fluidity can be seen as playing an active part in the formation of a collective imaginary.
44 Heiser, Jörg, *Post-Internet-Art–Die Kunst der digitalen Eingeborenen*, January 2015. Retrieved April 8, 2016, from http://www.deutschlandfunk.de/post-internet-art-die-kunst-der-digitalen-eingeborenen.1184.de.html?dram:article_id=30414

Observing from inside the Drift: The Studio as a Flux Condenser

Esther Moñivas Mayor

Written with the support of Fundación Montemadrid, 2013
Fellowship for Graduate Courses in Universities and Colleges abroad.

Words, sounds and things

A range of diverse sounds produced by two fine threads of air go through both masses of hydrogel. Due to imperceptibly changing factors, at a certain point the sounds become high whistles; then, they slip through space as repetitive, foamy bubblings and muffled booms. They erupt as noisy burps, or suggest polyphonies of voices that sigh, laugh, and scream in the distance.

žrrrrrcccccçç   flrssslępr ˜˜
grrrruuiiiiiiiiiīð   ˜˜˜   cj jc   cj jc   jjtrrrrrrruuuäuułł
eaaaaaäññaaaaiiiiii   • • • • • • •   ffr   ˜˜   niiiieeaaaa ɉɉ[1]

There are various reasons to begin this brief essay on the Liquid Things project by talking about sounds: first of all, because they probably make up the least graspable liquid of this investigation. Their drifting, in constant mutation and dispersion, is the motor of a fruitful imagination that feeds back into itself throughout the creative process, stimulating its participants to gauge a spectrum of physical densities and interactions between materials in search of nuances and variations.

Secondly, from an epistemological perspective it's interesting that these sounds elude scientific description, while at the same time offer seldom-travelled semantic pathways for exploring the complex physical and conceptual dimensions of matter. Their propagation in the space of the laboratory invites us to reconsider the difficulty involved in any linguistic approach to materiality. Gaston Bachelard pointed out that matter itself–flowing water in this case–generates a natural language, a voice that speaks to us of its own movements as well as of the flow of our own thoughts: "liquidity speaks to us of the very desire of language."[2] For Yunchul Kim and Roman Kirschner, to designate the materials with which they work is an often unnecessary and even limiting act; they favor the dimensions that sensory experience can construct on the scale of the material imagination.[3]

It's understandable that artists can move comfortably in a pre-linguistic, empathetic and direct relationship with the materiality with which they exchange information and energy as they work. However, how does one deal with the article, the catalogue, and, in general, any form of epistemological analysis of artistic materials and processes? What are the possibilities for using words to suggest the material essence of things and the nuances of the imagination that germinate around it, outside of that *natural language* that Bachelard alluded to? How can we name the systems of relationships and the events that the physical world and the imagination of the artist share?[4] The neomaterialist debates and the antimaterialist assumptions that have influenced the history of artistic media and materials in recent years take on a special significance as a background for the project at hand. As Coole and Frost (2010) pointed out, one of the paradoxes raised by attempts at a theoretical approach to matter is the unbridgeable gap that separates us from it in this exercise;[5] a space in which the immateriality of metaphors, meanings, interpretations and imaginations emerges as our distance from the material things themselves increases.

It could have been tempting to write this essay only with phonemes that take us to liquid materiality, but slipping between words and using them to attempt a theoretical approach to material processes presents a stimulating challenge. In our case, this narrative has centered on a concrete space-time system and has used a methodology that warrants explanation.

Inside the drift

There are important reasons to undertake internalistic approaches and direct, physical links with the spaces where materials are located and where they are transformed, even at the risk of losing the benefits of other forms of historical-artistic exegesis. To begin with, the elimination of the physical and temporal distance between the History of Art and its object of study allows it to document the conceptual and technical strata that build up during the processes of material experimentation. It also provides a record of the influence of environmental and relational factors in the creation of these complex semantic systems. Finally, it allows for the analysis of the material culture of a given field, and to witness the intellectual processes that surround the selection of substances, the handling of tools, the management of information and the establishment of relationships. Despite the approaches favored in recent years by the so called *material turn*,[6] there is still a notable lack of studies dealing with issues of materiality in art through an immersion in processes and a desire for bedraggling with the material itself.

One of the most lucid aspects of the Liquid Things project is precisely that it faces that lack, favoring the intersection of the processes of material experimentation with a theoretical meta-reflection on them.[7] Starting from a standpoint half-way between the History of Art and the technical knowledge of artistic media and materials, my own research joined forces with the open, experimental orientation of the project, forming an osmotic relationship with the practices of Roman Kirschner, Yunchul Kim and Karmen Franinović between January and October of 2014.[8] During this period, the artistic processes and the analysis of those processes took on ambiguous contours and experienced mutual contaminations.

Among the repertoire of strategies developed during this period it is worth pointing out the following: a) the testing of different points of observation within the studio; b) research into the theoretical sources and work methods of the artists; c) experimentation with various formats of scientific documentation (from a field notebook and sketches to photo and video documentation of the processes and their results); and d) active involvement in certain processes. A disciplined interrelationship of meetings and mix-ups, long periods of waiting, meals and walks, led to a mutual understanding deep enough to allow us to contextualize the different contributions to the project.

If these details are of epistemological interest, it's above all to show the paradoxes presented by the unstable balance between the scientific nature of research, and the influence of fluids on it. As Merleau-Ponty suggested, looking through liquids impregnates the gaze, deforms geometric structures and relativizes any kind of precision.[9] Taking this assumption as a point of departure, my goal was to observe the movements and the imagination of the artists *through the materials*. Because of their experimental character, the results present limitations, but we nevertheless feel that it was worthwhile to trace the network of relationships between particles, operations and dreams in this essay.

In any case, the relative position of molecules within a fluid affects its perception to such an extent that the time factor becomes the epicenter of most epistemological problems. The fluctuation of rhythms, the unpredictability of movements, the need to accept open results or even the absence of an end to processes… all of these things both influence the way an artistic practice is conceptualized, and also define certain conditions in the analysis of said practice.

If, for example, we take Deborah's number into account, even the *extended periods* of historical-artistic research suggested by Elkins (2008) are proven to be insufficient.[10] Said magnitude shows that the deformation of rheological behaviors is none other than a measure of the scale of time of the observation of this phenomenon.[11] From its formula it can be derived that the normal times of observation of the historian tend to make a solid of any substance. The paradigmatic case of glass – a material often considered an amorphous solid or a supercooled

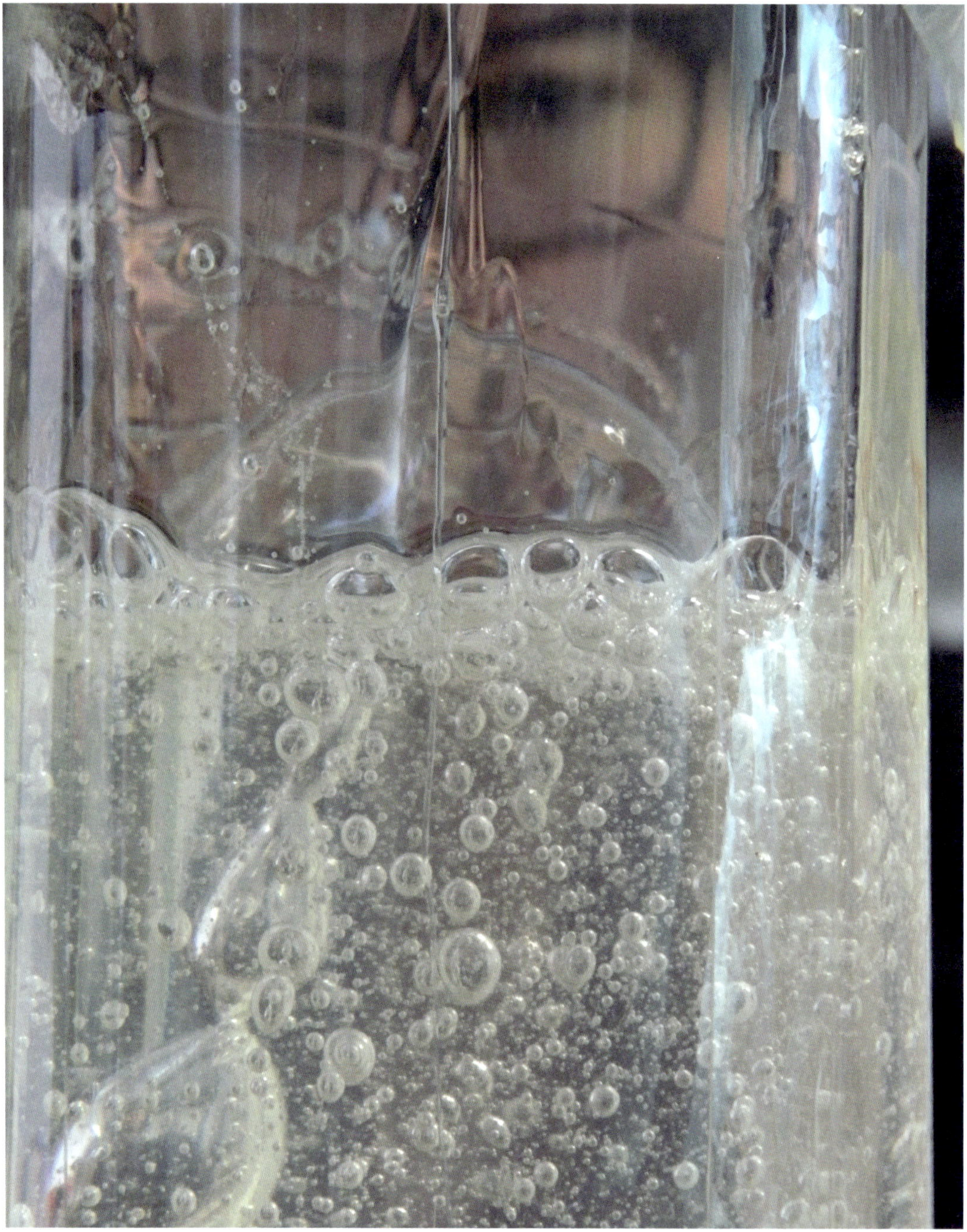

Fig.1: Yunchul Kim observing air moving through a PVA-polymer (Liquid Portrait Series)

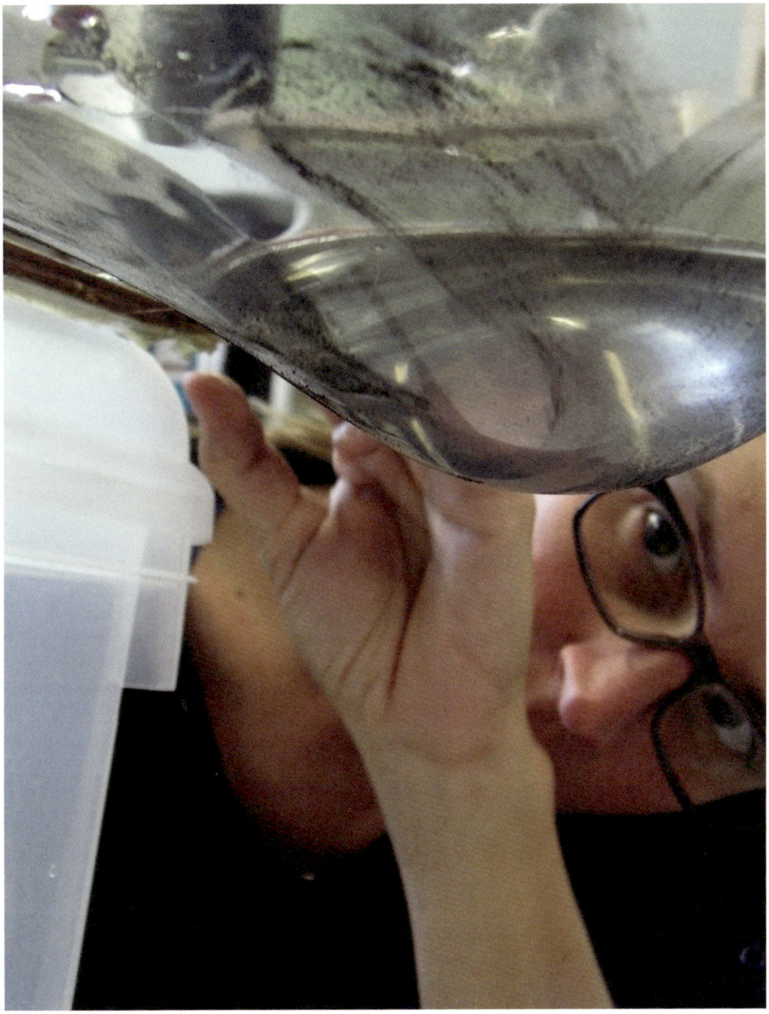

Fig. 3: Karmen Franinović experimenting with water and EAPs (Liquid Portrait Series)

liquid of infinite viscosity – shows this limitation in the normal epistemological considerations of the History of Art with respect to the solid and the liquid, the stable and the unstable. To give a notable example, the shifting due to gravity of material in the stained-glass windows of the cathedral of Chartres should not only invite us to relativize these classifications, but also to question the philosophical, political, and economic dimensions that the permanent flux of particles poses to concepts like *object*, *memory or patrimony*.

As a result, the study of fluid processes and imaginations *must itself be fluid* and incorporate within itself the properties of this state. This has direct consequences on the time-frame of the organization of the research – causing extended states of apparent latency as well as states of torrential activity – but it likely also affects more subtle aspects, like the kinds of movements that take place within the laboratory, methodological hesitancy or the emotional states of the individuals involved. To know *the fluid* is in many cases a question of learning to look and listen from the blurry margins of our rationality. It requires us to viscously and malleably perceive preconceived ideas about our senses, the names of things and their location in the cultural system.[12]

The flooded laboratory

To understand the space where Liquid Things took place is, to a large degree, to understand a fluid. Referring back to the basic definition, fluids are continuous media or substances whose low intermolecular cohesion leads to a relative position of molecules. In addition, any tangential tension causes them to be permanently deformed, regardless of the magnitude of that tension.

If we apply the basic parameters of a fluid to the ensemble of this project, we might ask ourselves the following questions: What tensions have led to transformations? Where, or based on what factors, should the density of the project be measured? What could be said of the viscosity, compressibility, durability etc. of its participants as catalysts of processes?

Inspired by the processes of self-organization described by Manuel De Landa,[13] at the most irrational margins of epistemic research a sort of observation of human agents as material substances is born.[14] Despite the risks involved in such a subjective perspective, this becomes coherent within the aforementioned theoretical framework. Thus, for example, certain movements of Roman Kirschner

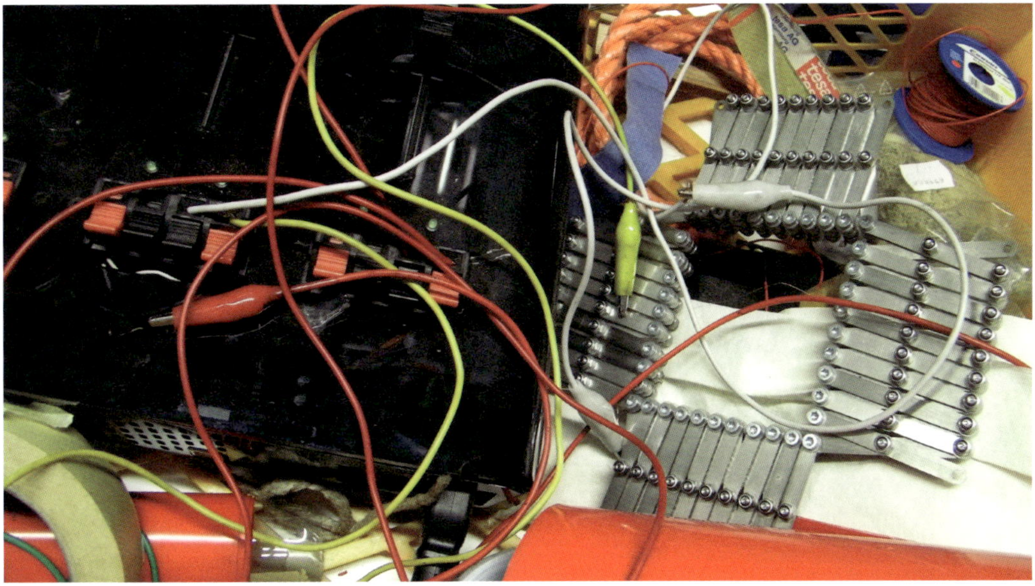

Fig. 2: Detail of a work desk in Liquid Thing's studio

can be perceived as a highly flexible semisolid. It's possible to observe certain agglutinative and sedimentary properties in Karmen Franinović, or Yunchul Kim's ability to act as a suspension formed by a solid phase of electro-active particles and a liquid phase with a high evaporation temperature.

With regard to the density of the project, it's clear that the right place to perform this measurement is at the studio-laboratory, where the processes take place and where the liquid things evolve. The luminous room, at the end of endless deserted corridors that isolate it from the din of *Wien Mitte – The Mall*, presents itself as a communicative space where processes emerge, interferences occur, reading takes place, and snacks are sometimes had. At times, the air seems soaked with emanations from the building itself mixed with substances from the laboratory. A faded red couch and a table with sugar cubes invites us to sit down and begin a conversation. Machines, tools, documents, substances and dreams are interwoven over five or six tables, located as to favor a flow of simple movements. When the artists move around the lab, their movements are in fact gentle and not easy to perceive.

It would be impossible to describe the entire spectrum of materials used in this space. Among them, some have a privileged place because of the more intense use that is made of them – this is the case of certain oils, silicones, hydrogels, polymers, magnets, electro-active substances and meta-materials.

The floor and the surfaces of the tables are fields where amalgamated components, rather than defined entities, relate with one another in a variety of ways. These miscellanies develop over time through the physical-chemical reactions they undergo, and also through the semantic games suggested by their syntax. Cables, screws, polymers and tools are tangled and tarnished together, occupying all the available space. Their relative positions vary as the weeks go by, but a certain cohesion between the components favors the maintenance of constellations.

On one occasion I performed the experiment of rearranging the components of a table in a different way, leaving a large, empty space in the center. I placed the portable electric burner and the electric scale on one of the long sides, and on the other one, two rows formed by all the bottles and small beakers, moved there in the same order in which they were found. Puzzled, Kirschner asked about the reason for this change in the delicate system. In less than two days the central area was again filled, with a distribution of elements reflecting each of the actions performed on the table. This allowed me to observe how that arrangement was slowly reached with each movement, but also how spills and notes on quantities written in pencil gradually enriched the table.

One of the most stable constellations agglutinates elements linked to practices with electro-active polymers (EAP), and therefore with electrical processes and the deformation of ultraflexible sheets.[15] Here we found a handmade electronic device used to work with EAP, a jar of graphite powder, a roll of double-sided acrylic VHB, wires, frames of various sizes and a curious mechanism used to stretch VHB tape in two directions. In the hands of Karmen Franinović the film is transformed into a fine, mobile membrane able to hold a certain amount of water on one of its faces. The graphite powder spread on the bottom side of the film allows areas to be defined where a magnetic field, applied by electrodes, deforms and moves the liquid in sequences and intensities that can be modified by a controller. It's easy to forget the dangers of combining water and electricity in these experiments when faced with the haptic and visual pleasure suggested by these membranes containing water, deformed and undulating rhythmically by way of electrical impulses, as though they were the skin of jellyfish. Electricity and water are in this case two symbiotic elements that combine to move these membranes and our psyche.

Distributed at strategic points in the studio, another constellation formed by devices invented or produced by Kim and Kirschner, calls to mind the work of

Panamarenko. Rather than taking off from the ground, these machines aspire to form systems that automatically produce evolving flows of forms and sounds, at the same time that they come together in different processes. From one corner comes the pounding rhythm of a 3D printer made with a DIY kit. A little farther away lie a couple of Arduino boards and a group of multicolored cables, along with a pH controller waiting to be finished. Next to the entrance a curious device purrs: made of concentric rotatory wooden frames powered by electricity, it causes liquids to move circularly in three dimensions. Various transparent spheres suspended within the frames serve as incubators for the ultra-slow processes of polymerization that the various fluids undergo. The progressive viscosity of these fluids is tested by measuring their friction with small elements–tubes and nail heads–which, as happens with stalactites, cause deposits of the substances to be formed. Kim and Kirschner test different patterns of movement for this device and try to adjust the density of its sound, hanging pieces of *silly putty* at different points to compensate for the weight, and also probably to motivate new experiments with silicone-derived substances.

Magnets are another element in this studio-laboratory that create a web of invisible relationships. Cubic or spherical in shape and of a variety of sizes, they can be found discretely attached to table lamps, metallic walls and shelves, like little insects that take the eye some time to identify. Made from a mixture of neodymium, iron and boron, these super-strong magnets generate their own incorporeal map of the activity within the laboratory. Kim considers this beneficial, despite repeated experiences with data loss in electronic devices. Due to their extremely high remanence, some magnets are very difficult to remove from the places where they are located, and cause unexpected movements when a metallic object comes near their field. It could seem that they are there to *trap* sketches or ideas, but they are actually the protagonists in a dance of suspended particles that suggest waves, whirlwinds and complex formations of liquid planetary bodies in certain of Kim's pieces, which one can't help but think of in *solaristic* terms.[16] In them, he experiments with the rheological and visual properties of certain metamaterials that he himself creates,[17] playing with the propagation of light and color inside them; with the response to magnetic fields; or with the effect of gravitational forces. The magnets, therefore, suggest complex, multi-step processes involving other substances, like paramagnetic nanoparticles, photonic crystals, colloidal suspensions or pigments.

The walls are the space where concepts, maps, names, and sketches collected by the artists preferably accumulate and float, meaningfully suspended by

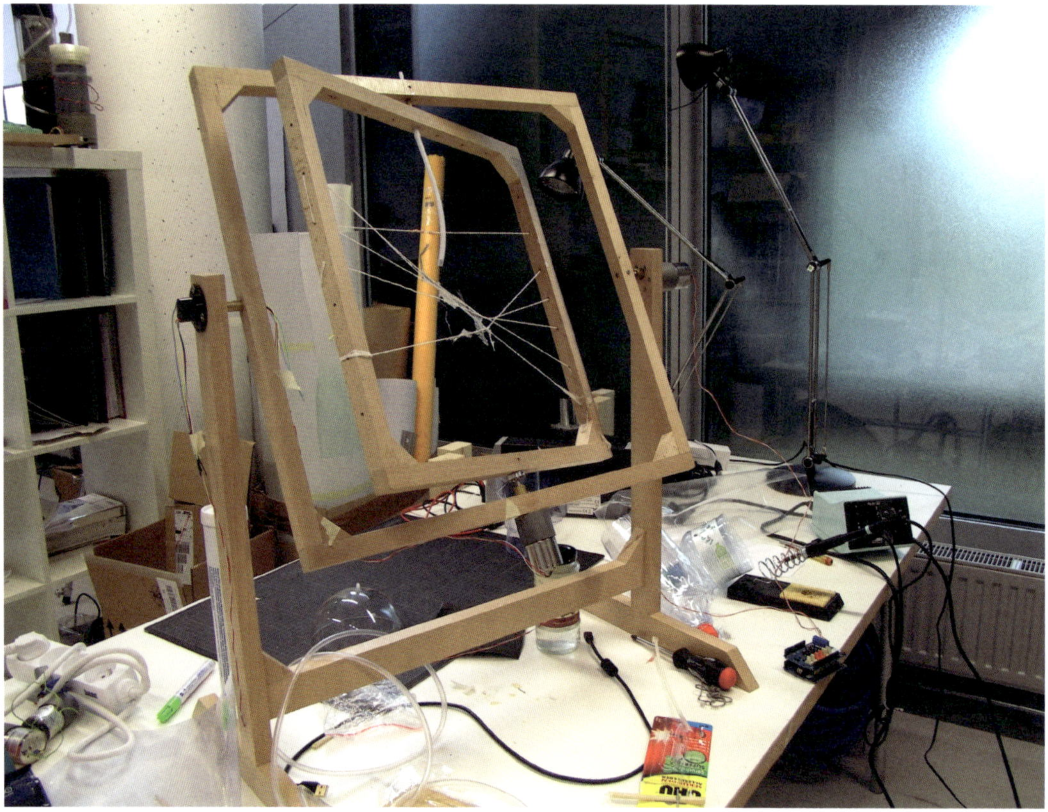

Fig. 4: Detail of a work desk in Liquid Thing's studio

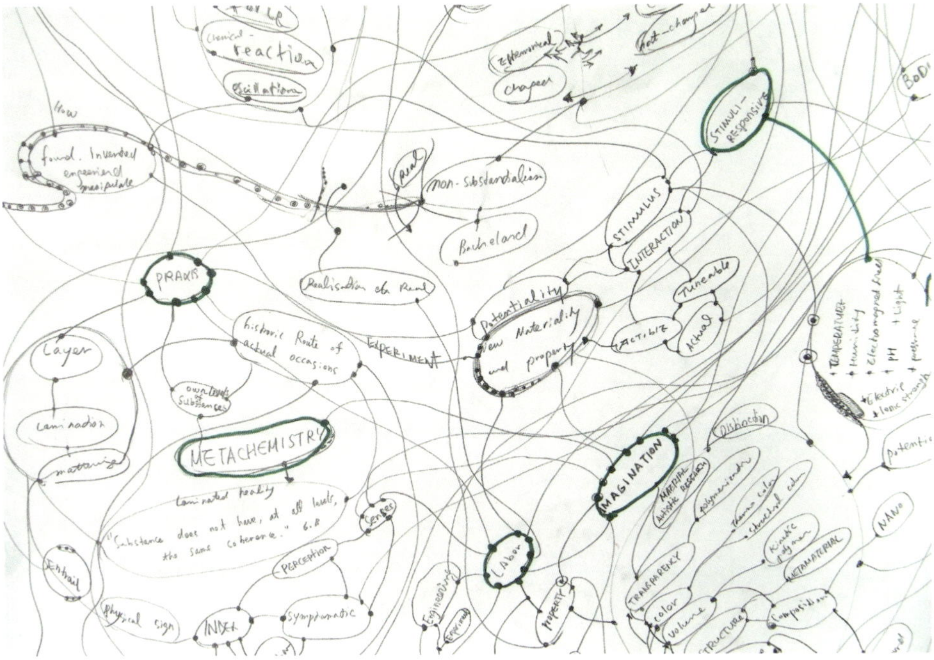

Fig. 5: Detail of Kim's conceptual maps.

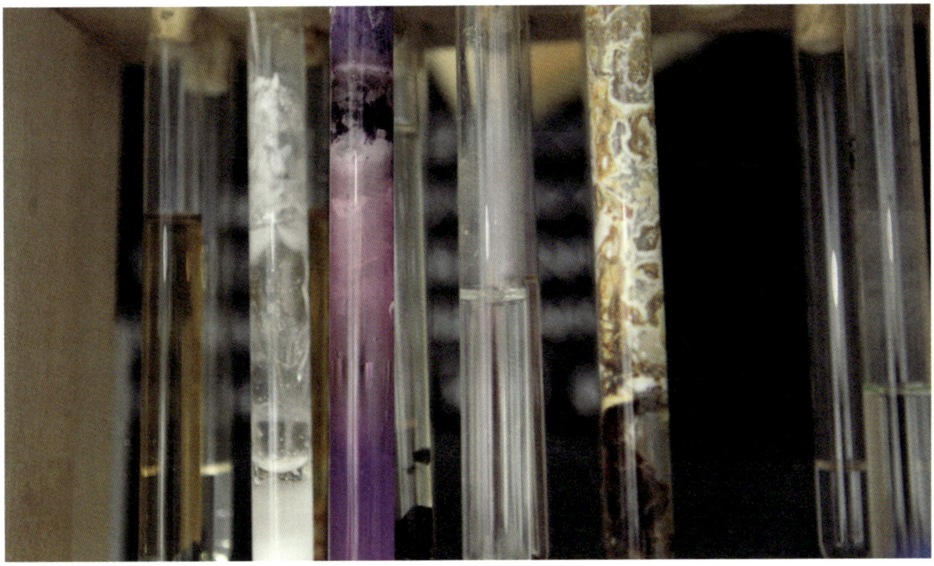

Fig. 6: Detail of test tubes

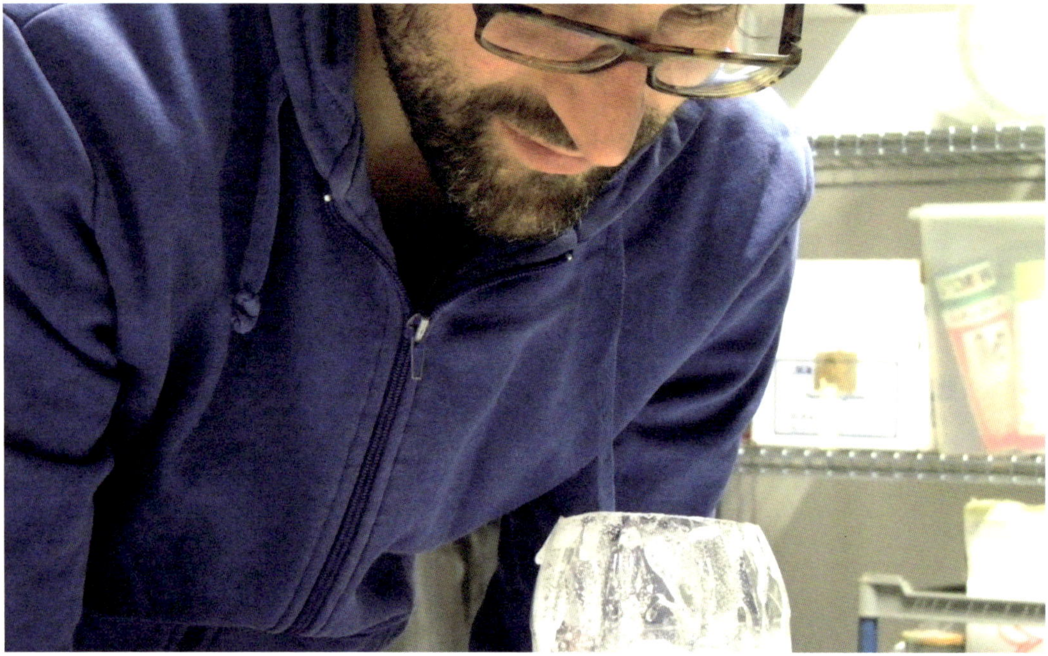

Fig. 7: Roman Kirschner observing the kinetics of a PVA-polymer (Liquid Portrait Series)

those magnetic forces. Conceptual maps make evident other tensions that have transformed the environment, added to electromagnetism, humidity or gravity. Lying on the tables and arranged on the shelves are documents and administrative papers about the project. Printed scientific articles, books and catalogues also move between different horizontal and vertical surfaces, and on occasions singular names are preserved on the maps on the walls. Gaston Bachelard, Lambros Malafouris, Bruno Latour, Joseph Earley, Karen Barad, Tim Ingold, Stanislaw Lem, William Wordsworth, Ilya Prigogine, Alfred Nordmann, Erik Satie, Alfred North Whitehead, Michel Serres, Peter Blegvad, Mircea Eliade, Paul Needham, Novalis… They appear and disappear on these surfaces, fused with the tangential forces that other researchers have marked the project with in workshops, symposiums or conversations.

Also surrounding the room, the shelves filled with boxes allow the tables and floor to be supplied with new components that, only circumstantially, come back in the form of classified tests. This is the case of the processes of chemical oscillation and other surprising changes in the materials that have been tested.[18] Like a collection of indescribable landscapes, the Petri dishes and test tubes organized in racks contain the remnants of moments of aesthetic and sensory fascination. Various containers of molecular kitchen products lie as a reminder and the ephemeral production of spherifications and closed, permeable membranes which Kirschner and Kim use to experiment with form, movement, osmotic pressure or permeability. Transparent, colorless glass can be found all over the shelves, tables and floor: jars, funnels, beakers or agitators ready to be used at any time.

In the end, the areas of the lab most active during my stay were the ones linked with the production of hydrogels and bubble machines that inevitably call to mind the experiments of David Medalla in the 1960's. Kim and Kirschner's research delved into determining the levels of optimal viscosity of different mixes considering the mobility of the particles and the speed of evaporation; exploring the resistance of containers of different shapes and materials; observing patterns of movement created by the movement of air in the hydrogels; and testing the sonic qualities of these mixtures.

The conjunction of water, polyvinyl alcohol and boron salts initially led this experiment toward the creation of a non-toxic compound whose viscosity could give interesting results when air bubbles passed through it. Erratic pathways, surprising trajectories, varied sounds, solidifications of foam into delicate laces were part of this performance, but once again the research was enriched by a series of unexpected events.

Due to its movement, the liquid escapes from its containers: it drips onto the little electronic scale and the electric oven used to measure quantities and control the temperature during the mixing phase; it spills onto the tables because of how hard it is to handle; it bubbles out of the upper opening of the containers once the air pumps that animate the interior are installed. Finally, there are times when the plastic support structure falls clamorously under the weight of an ever increasing amount of material. At this point, the most interesting thing might be the way in which any of these accidents modifies the direction of the research. One day, Kirschner identifies an extremely light, white cloth on the surface of the floor, like a delicate spider web on the cement. This sort of fabric, the result of an involuntary spill, becomes a priority and opens a new line of research.

The delicate remains of a series of accidents hang by threads for all to see: ideas waiting to find their moment or their role within the research, and also beacons of that same research.

Without a precision scale

In general, it could be said that the activity of the Liquid Things laboratory offers nothing that could be considered stable. The general research methodology reveals a network of complementary actions, only partially fixed in terms of their development over time. The absence of white lab coats and precision scales shows a desire for distance from smartness, asepsis and exactness in order to let in surprise, play and reflection. The epistemological analysis of artistic processes also requires a methodology adapted to the task of combining scientific documentation and detailed knowledge of the tools and material involved, with prospecting in the difficult to reach field of imagination.

The more we look into these materials and processes, the more we lose the ability to say something concrete and definite about them. The ontology of the substances dissolves as we become aware of the layers of meanings that form our gaze, showing the uselessness of the desire to catch them under one name, one formula or one container. We can just maybe find temporal coagulations of meanings that, in any case, always lead us far beyond what is before us. The epistemology of processes becomes increasingly complex as different dimensions are taken into account, dodging the limits of the presupposed meanings that must be involved in said analysis. In the case of fluids, there is also a certain negation of rational dimensions; this makes all analysis slightly blurry, as if in a permanent embryonic state.

Nevertheless, we might say that the whole laboratory, with its occupants and its dreams, is the flowing compound in which processes are born. This intimate space, with no rules or pre-established readings, is a condenser of flows and mutations that resists being trapped in images and narratives, but that still constitutes a catalyst of imagination, experimentation and knowledge. One therefore understands that liquid substances interrelate in a physical, philosophical and political order in which reductionism derived from control and from hierarchical exercise of power has been substituted by a discipline of emergent processes.

1 Transcription made by the author of a real sequence of sounds.
2 Bachelard, Gaston, *El agua y los sueños. Ensayo sobre la imaginación de la materia*, FCE 2002, p. 279
3 Conversation with Yunchul Kim and Roman Kirschner, April, 2014. Here we take the concept of *material imagination* advocated by Gaston Bachelard.
4 The term *event* is borrowed from the theories of the philosopher Alfred North Whitehead (1861-1947), who criticized traditional categories of philosophy for their failure to convey the essential interrelation of matter, space, and time. In his *Process Philosophy*, Whitehead proposed to understand the objects as a series of events and processes. Whitehead, Alfred North, *Process and reality. An Essay in Cosmology*, The Free Press, 1929
5 Coole, Diana and Frost, Samantha (eds.), *New Materialisms, Onthology, Agency and Politics*, Duke University Press, 2010
6 See, for example, Bolt, B. and Barrett, E.: *Carnal Knowledge. Towards a New Materialism through the Arts*, I.B Tauris 2013
7 See *Liquid Things (Modules)*, retrieved Aug. 12, 2015, from http://www.liquidthings.net/?page_id=662
8 Moñivas Mayor, Esther, *Art history integration in transdisciplinary art and science university programs. Field research on creative processes with fluid materials*, Research project developed between January and October of 2014 in the University of Applied Arts Vienna (Art and Science Department, project Liquid Things), funded by Fundación Montemadrid.
9 "When through the water's thickness I see the tiled bottom of the pool, I do not see it despite the water and the reflections; I see it through them and because of them. If there were no distortions, no ripples of sunlight, if I saw, without this flesh, the geometry of the tiles, then I would stop seeing the tiled bottom as it is, where it is, namely, farther away than any identical place. I cannot say that the water itself–the aqueous power, the syrupy and shimmering element–is *in* space; all this is not somewhere else either, but it is not in the pool. It dwells in it, is materialized there, yet it is not contained there; and if I lift my eyes toward the screen of cypresses where the web of reflections plays, I must recognize that the water visits it as well, or at least sends out to it its active and living essence". Merleau-Ponty, Maurice, *Eye and Mind*, Northwestern University Press, 2007 (1964), p. 371
10 Elkins, James, *On Some Limits of Materiality in Art History*, in 31: *Das Magazin des Instituts für Theorie*, Vol. 12, Zürich 2008, pp. 25-30. (Special issue *Taktilität: Sinneserfahrung als Grenzerfahrung*, edited by Stefan Neuner and Julia Gelshorn)
11 $De=\lambda/t$ where λ is the time of relaxation and t is the time of observation. Solids have an infinite time of relaxation, while in liquids this value nears zero.
12 This is an inexplicit reference to Michel Foucault's introduction to *The Order of Things*. Foucault, Michel, *Las palabras y las cosas: una arqueología de las ciencias humanas*, Siglo XXI 2002 (1966)
13 De Landa, Manuel, *A Thousand Years of Nonlinear History*, Swerve Editions 2000
14 These descriptions only refer to the period indicated.
15 EAPs are polymers that exhibit a change in size or shape when stimulated by an electric field. They usually undergo a large amount of deformation while sustaining large forces.
16 Here we make a reference to the extraordinary oceanic activity of the planet Solaris, protagonist of the science fiction novel written by Stanislaw Lem. Lem, Stanislaw, *Solaris*, Minotauro 2007 (1961)
17 Highly complex artificial materials.
18 Chemical oscillations are bidirectional reactions. Depending on the mixture of chemicals, sequence of colors changes, consecutive emissions of gas or precipitations occur, depending on the decrease of free energy in the mixtures.

Sources of the figures:

Fig. 1-7: Photographies by Esther Moñivas

Thinking Active Materials: Actively Thinking Materials

Karmen Franinović

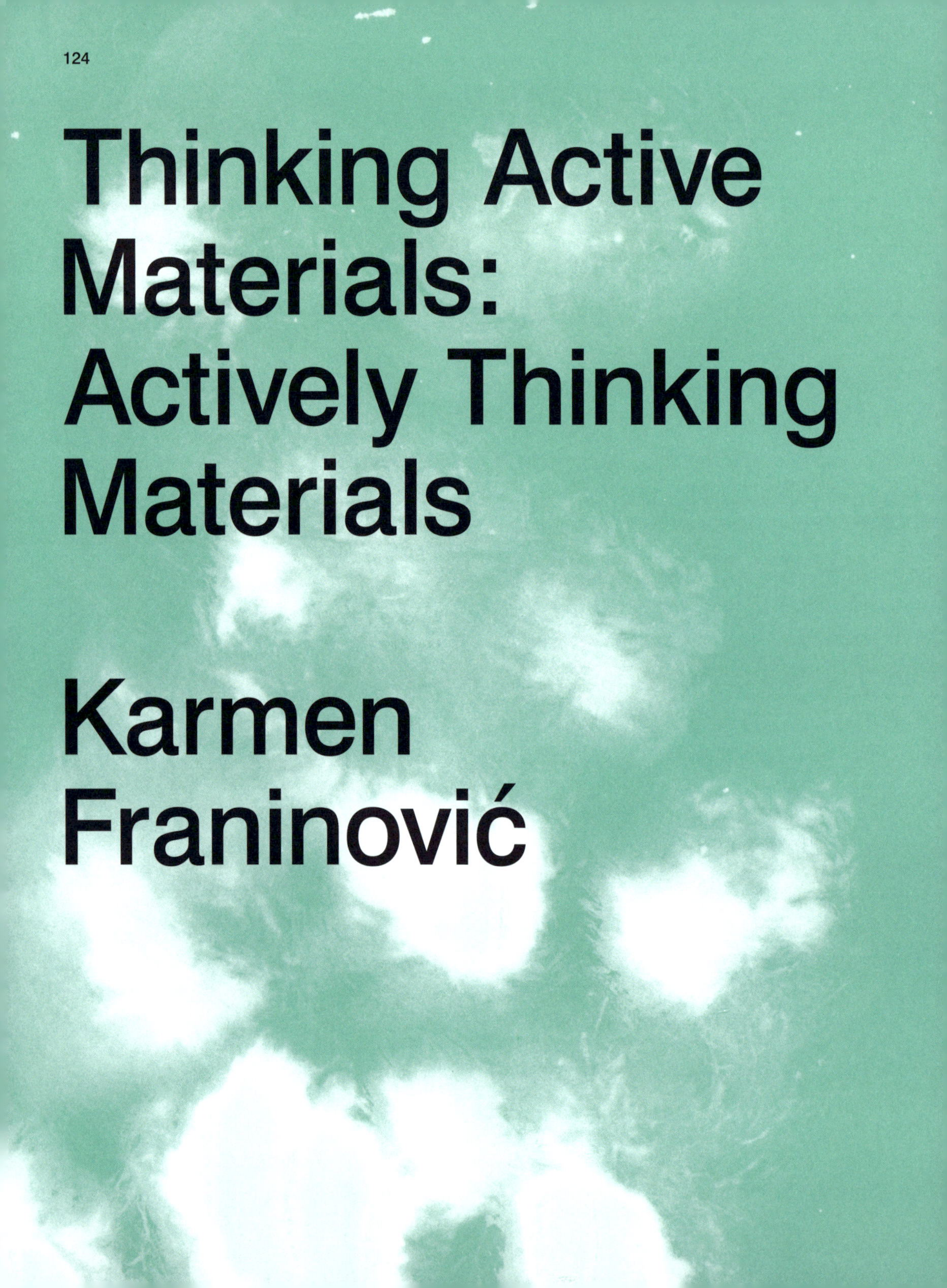

A not-yet-known material

There is something exciting and at the same time uncanny about holding a novel material in ones hands. There is no expert to turn to, no technique to follow, no crafting skill to receive. At first encounter, there seems to be nothing to compare a novel material to, nothing to tell us what its purpose may be or how to approach it. Its qualities and processes, however, pull us into relation. They invite us to sense, to feel and to manipulate. Events emerge through our mutual engagement, "as (the sculpture) gently swings the liquid around its surface, we begin to move in sync with each other [...]".[1] Moved by and attending to the activity of a material, we learn to know what a material can do.

In the Enactive Environments research group, we try to stay close to such transformative experiences when thinking about and with active materials. The perceived stability of matter, existing research methods and even mental

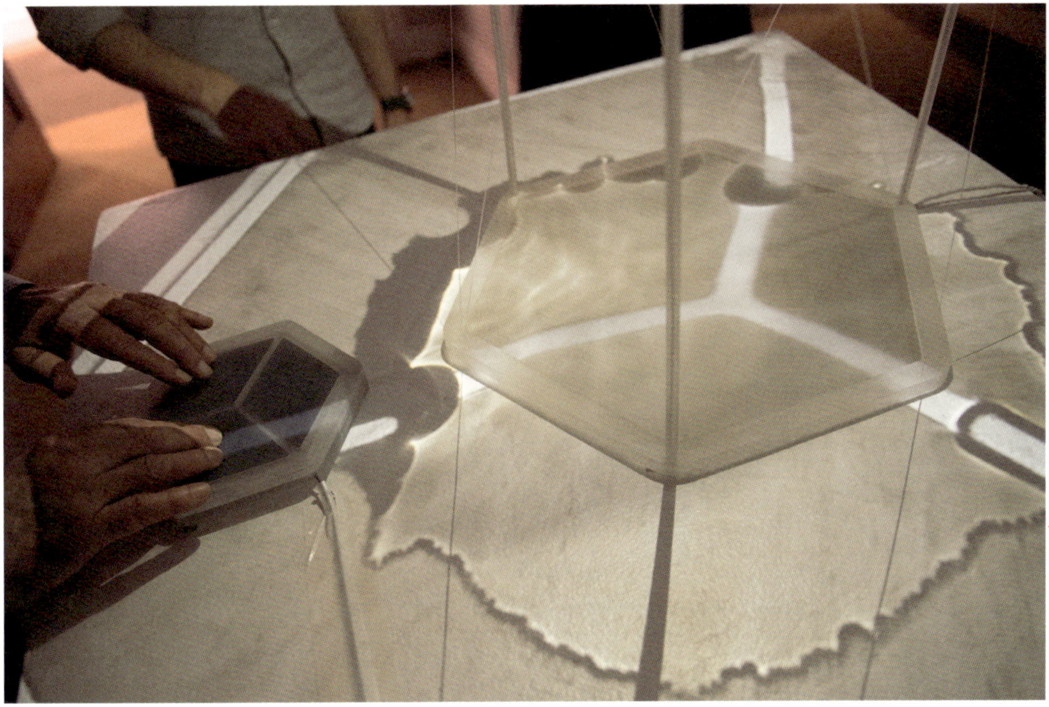

Fig. 1: Visitors interacting with a shape-changing sculpture in the exhibition *Fluid Morphologies*.

associations may push us into reducing and freezing the potential of a dynamic material. When we objectify matter by studying its specific states but neglecting its transformative power and the ungraspable in-between states, our thoughts risk becoming equally fixed. A not-yet-known dynamic material may help us go around this issue – because its processes are unstable and influenced by its surroundings, but also because there is not much that connects it to something already known.

Escaping the stable mental associations of a novel material, such as references to nature or animal life, has become an essential criterion for our work. If people perceive always the same object in a particular formation or process, there would seem to be something wrong with the way we constrained a specific material. We blocked and froze its potential of becoming into a fixed image, a fixed thought. Its vitality has been transferred onto a reference from the past, something already predetermined and stable, rather than actively opening it to the present. So how can we instead keep the vitality of matter and thought in the here and now? How can we extend our lingering in the unfamiliar and explore the not-yet-known? How can we enable flows of materials and thoughts that escape the solidification into objects and ideas? How can we think materials actively and how can we enable the activity of materials? These were the central questions of *Material Aktiv Denken*, the participatory event discussed in this essay.

Active Materials

Often presented through scientific formulas describing their performance, novel materials are either produced in engineering contexts oriented towards purposeful applications or in basic research exploring their processes and properties. It is the latter perspective, specifically on inherently dynamic materials, that is the focus of our research. However, our work is conducted in an artistic context and, while informed by scientific findings, it is not driven by any kind of purposeful application or scientific proof. At the same time, our research engages with discourses around new materialisms, which largely do not incorporate direct engagement with materials in their research process. As Karen Barad herself put it,

> *Language matters. Discourse matters. Culture matters. There is an important sense in which the only thing that doesn't seem to matter anymore is matter.*[2]

Active Materials is a working concept that describes our approach to inherently dynamic properties of materials, which sets to uncover their aesthetico-political potential. Considering the dynamic nature of such materials, when we talk about a material, we also refer to its transformative processes and events. Following Brian Massumi's activist philosophy, the aesthetic does not stand here for a reflective and distant contemplation of the world, but for self-enjoyment in the midst of active self-creation. This quality is not exclusively human, but also includes *dumb matter* which can be experientially self-creative. Co-existent with the aesthetic aspects of a process, are the political ones, characterised by relational and participatory qualities. As Massumi wrote, "These aspects are not treated as in contradiction or opposition, but as co-occurring dimensions of every event's relaying of formative potential".[3] In our research, we examine how these aspects unfold together over the course of our research activities such as the *Material Aktiv Denken* workshop.

Materials we work with are inherently capable of changing their states or properties in continuous and complex ways. For example, they may change shape and viscosity in different environmental circumstances, or they may glow and sound when exposed to changes in light, temperature, electrical charge or pressure. Such materials are often labeled *smart* to reflect their efficiency for a specific purpose defined a priori by humans. Contrasting such an approach, in the Enactive Environments group, we invent new processes of working with novel materials, in order to take them outside of scientific laboratories and into the hands of artists and designers.[4] We embrace the disobedience of materials, and work with *errors* and qualities that are undesired in engineering contexts.[5] We let the unexpectedness of a material process guide our research process. We search for ways of opening up alternative paths for novel materials, to let them leak into the world and affect our imagination and thought.

Together with the Liquid Things group, we set out to explore some of these ways in a participatory context. *Material Aktiv Denken* was a thinking-making event on active materials that took place over four days at Zurich University of the Arts.[6] The goal of the workshop was twofold. On the one hand, we aimed for a thorough reflection about active materials and their unique behaviours and performances. On the other hand, we hoped to advance critical and artistic approaches that conceptualize materials and materiality in a more active and performative way. The meaning of the title Material Aktiv Denken reflects this double goal. Its translation to English could be, *Thinking Active Materials* as well as, *Thinking Materials Actively.* When thinking about active materials we ask: What makes a material active? How can we work with inherent activity of materials in an aesthetic and relational rather than a functional way? What kinds of events can be teased out of novel materials? When thinking materials actively, we wonder: How do active material properties and processes interact with concepts and theories of new materialism? How does material engagement activate our imaginations and thoughts? How can we create contexts to think materials actively?

Within the *Material Aktiv Denken* workshop, we aimed to create a collaborative space in which we could playfully, yet systematically, shift between materials and concepts. We searched for ways of tackling materials and concepts through workshop activities in which they could quickly manifest themselves to then dissolve or merge into another thing or thought. In this interplay between thinking and doing, we wanted people from different backgrounds to experientially approach activity, process, transformation and flow. To confront these many challenges we started by considering different spatial and temporal conditions that could foster emergent activities.

Seed-Spaces

> *I am in favour of indeterminacy and against automatic planning. Against any planning, qua planning, as a shifting to the future of what is being done now […] To plan is to deny action and the present their interest and reality […].*[7]

How can we prepare an event without overly predetermining it? Focused on people rather than on materials, my work explores this question in urban contexts. By means of interactive installations, citizens are invited to interact and change their surroundings. These *seed-spaces* are spatial interventions composed of physical, both tangible and non-tangible, conditions which, while visually provoking, lay latent until intentionally activated by passersby. Under their influence, *seed-spaces* begin a physical transformation of urban location while enabling unexpected social and sensorial experiences.[8]

The origins of this approach can be found in early interactive architecture and cybernetics, in so-called strategies of underspecification. By underspecifying the goal of an architectural system, it was hoped that it would evolve through the conversation with its inhabitants and thus escape a problem of a functionally fixed, purpose-oriented machine. Gordon Pask, a cybernetician and artist, argued that an architect should, "provide a set of constraints that allow for certain, presumably desirable, modes of evolution [...]".[9] These constraints would then have allowed the system to evolve on its own, to predict the needs of the inhabitants and to transform accordingly.

Similarly, a *seed-space* is a set of constraints or conditions and uses interactive technologies, but its goal is not to predict inhabitants' desires or needs. Rather, the *seed-space* flirts with existing behaviors to produce new ones while revealing something about the qualities of a location and its inhabitants. It shows how interactive technological apparatuses influence a specific situation, and are permanently open to change and new iterations. The often-surprising discoveries are based on direct and participatory observation, which occurs after setting the seeds in a specific location and allowing them time to evolve.[10]

Transferring *seed-space* techniques to material research may help us to not overly predefine material processes but allow them to emerge, offering us surprising insights. However, while the urban seeds were placed there for people to generate events, active materials required different conditions to enable their own processes to emerge. Thus, for the *Material Aktiv Denken* workshop, we had a double-sided challenge: activating materials themselves and activating people's thoughts and actions with and through materials. So what kinds of conditions and constraints can help us achieve this goal in a workshop setting? How can people and materials share the same seeds? Can materials be seeds for people, and the other way around: can people be seeds for materials? How does the status of a human participant change in this context?

Seeding

Our preparations for *Material Aktiv Denken* adventure started by selecting concepts and authors which we felt were critical for a stronger connection between the practice with active materials and the theories of new materialism. The two research groups had previously collaborated in the workshop in which we experimented with self-expanding and self-contracting matter.[11] We worked with electroactive polymers which change shape and size when exposed to electrical fields and electrorheological fluids which have the capacity to reversibly change their consistency from an elastic to a hard gel when exposed to electrical fields. Particularly salient in our exchange were the notions of process, material/ity, agency, activity/passivity, affordance, material engagement, enaction and transformation. We chose authors whose writings resonated with these concepts, and discussed how they could connect to different materials we were working with. Our workshop reader included texts from K. Barad, J. Bennett, M. DeLanda, A. Gell, J. J. Gibson, T. Ingold, L. Malafouris, B. Massumi, J. Soentgen and H. P. Hahn, A. Pickering and P.-P. Verbeek.

In order to interweave texts and materials, we had to prepare seeds for material events to take place. The mapping of different concepts, materials and techniques allowed us to test some ideas before the event. From there, different experiments were planned choosing materials that fit the concepts explored and questions asked. These included dry ice, glowing sugar, chocolate clouds, spherified food, photonic crystals in oily suspension, electroluminescent paper and ingredients for Belousov–Zhabotinsky reactions. The timing of the material processes and the timing of the workshop activities had to be choreographed in advance. Based on selected readings and material experiments, the workshop activities were loosely structured by different themes: on the first day the focus was on agency, on the second day we engaged with materials from the perspective of human/non-human actors and related affordances; on the third day we approached the imagination and material transformation; and on the last day we reflected on the workshop.

We discussed how various participatory techniques could help us create a space of openness and trust between participants coming from different backgrounds. Participatory workshops can be understood as a kind of third space in which diverse parties produce shared knowledge.[12] Most techniques for creating such space are verbal and narrative, making storytelling one of the preferred methods. More recently, research has shown that a non-linguistic collaborative experience, especially through the use of physical artifacts, could help bridge the gap between people with different expertise.[13] However, this research used prop-like artifacts that can enable specific sonic interaction and the aesthetic qualities of the matter other than sonic remained under-explored. To circumvent some of these issues, we decided to modify the collaborative techniques, as described in the sections below.

We invited colleagues working with related topics from the fields of philosophy, art history, art, design, architecture, and material science to join the workshop. In advance of the event, they received the reader, and as a means to introduce themselves they were asked to bring an object and a material they considered to have active properties (in any sense from imaginary to physical). In the past, this kind of introduction allowed us to shift the attention away from the skills and accomplishments of individual participants, while providing a personal angle to the topic of the workshop.[14]

Launching Things and Thoughts

The introduction through an object and a material, rather than participants' professional background, proved once again to be an excellent way to start a workshop. For this *Activating Object/Material* exercise, most participants brought one thing and presented it both as a material and an object. This often occurred through a transformation from a material to an object or the other way round. Paper was folded into different plane shapes and sent to fly over the room guided by their sharpened tips. A dry leaf was crumbled into dust and sprinkled over the table. The participants' introduction turned into an individual or collaborative performance, and the materials entered the stage as active and activating participants from the start (Fig.2).

Following this warm up, we approached the readings of Ingold, Gell and Gibson through a *Concept Speed Dating* exercise. In it, concepts were quickly

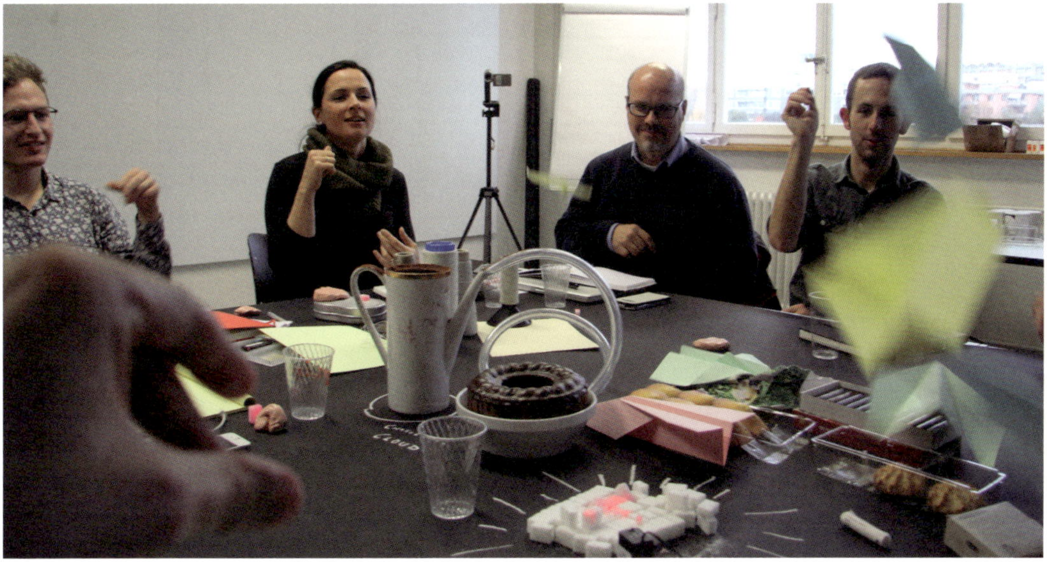

Fig. 2: Participants throwing paper planes: personal introduction through brought materials.

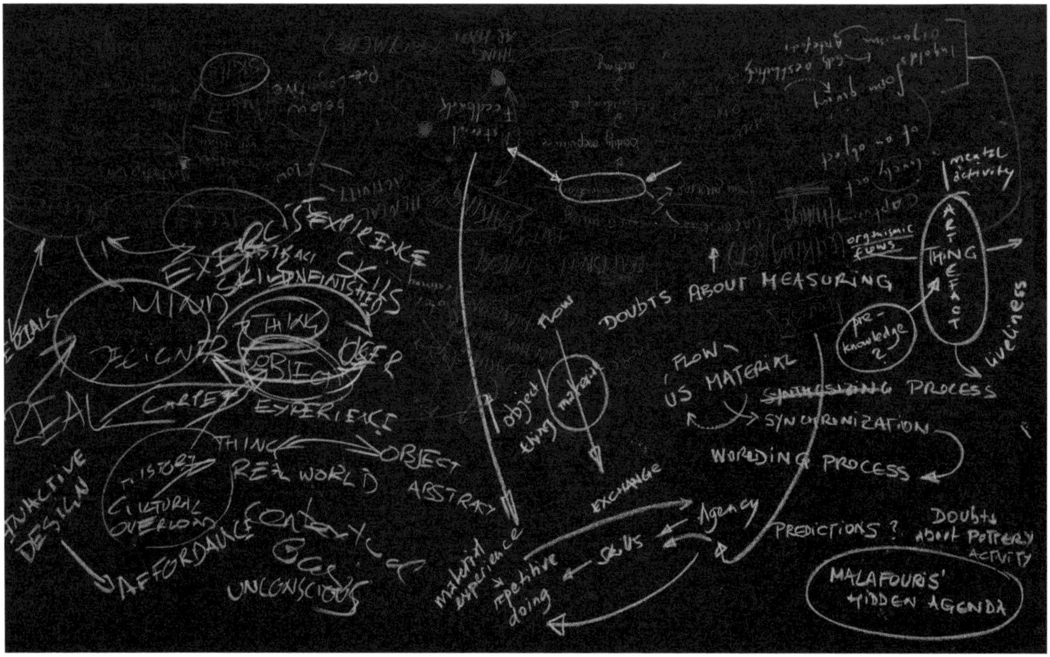

Fig. 3: Detail of black sketching paper with traces of the *Concept Speed Dating* showing different conceptual flows.

discussed in couples and after several minutes, partners were changed and another discussion started.[15] We extended the original technique by adding a specific spatial and recording arrangement: a long table on which the participants could draw or make diagrams as they spoke. On one side of the table, the participants used white pastel markers and remained seated when the couples were to change. The other side of the table was the mobile one, where each participant had a specific color that she would take with her as she moved to the next partner. Afterwards, the colors allowed us to trace how the people and concepts traveled during our discussions. As a frozen river full of liquid diagrams and concepts, this long sheet of paper helped us identify and discuss individual and collective flows of concepts (Fig 3).

The day ended with first demonstration using liquid materials: a spherification process used in molecular cuisine that Kirschner had worked with in the course of the project Liquid Things. He carefully encapsulated a liquid into blobs: as the red alcohol dropped from a syringe into the water it suddenly appeared solid,

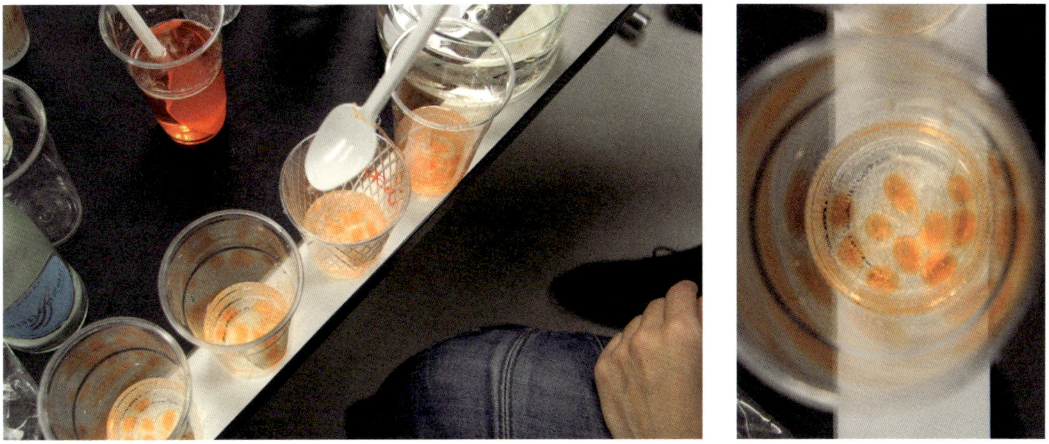

Fig. 4: Preparation of *Spherified Aperol*.

dividing into separate spheres (Fig 4). The containers and the contained, these spheres could leak into their watery surroundings only over a long period of time. In the short run, they could be turned back into liquid by squeezing them. We ingested these little blobs, feeling their boundaries dissolve or break in our mouths. As we drunk this *Spherified Aperol,* we got a taste of the material processes happening in front of and within our bodies.

Fluid Affordances

On the second day, with the memory of spherified drinks still in our bodies, materials continued taking over the stage. We were examining how humans and environments can affect material processes and vice-versa. The question of what is active shifted towards how is it active. In the *Let your Material Act* exercise, participants were asked to show and talk about how the materials of their choice act. This is how one participant described his experience, "when you just touch it, it sounds a little bit like one of those plastic wrappings for a toy [...] for me its highly counterintuitive because it has counterintuitive sound qualities...and flow qualities because it's so slow".[16] The material was Silly Putty, a silicon based polymer, a type of non-Newtonian fluid, which acts as a viscous liquid over a long time period and as an elastic solid over a short time period. Not having a specific purpose, Silly Putty offered many ways of acting, which changed as its properties changed. This suggested that the affordances of a dynamic, and in this case purposeless, material might be different than those of an object, which for most people invites habitual ways of use.

Donald Norman proposed the currently dominant view of affordance linked to this kind of instrumental use. He defined a design affordance as "the perceived and actual properties of the thing, primarily those fundamental properties that determine just how the thing could possibly be used. A chair affords *(is for)* support, and, therefore, affords sitting".[17] The other possibilities that a chair may afford, such as rolling or kicking, are backgrounded by the function of the object, which is fixed by habitual use. However, the inventor of this concept, psychologist J.J. Gibson, had a more relational understanding of affordance. He writes, "I mean by it *something* that refers to both the environment and the animal in a way that no existing term does. It implies the complementarity of the animal and the environment".[18] For Gibson, the affordance was not solely a set of physical and perceptual properties of an animal and of an environment, but a potential for action, which emerged from their relation. Although biased towards animal and human perspective, Gibsonian affordance can be read as an offering of the environment, an invitation to explore the potential for novel action rather than to repeat the established one.

In perceiving an object at a distance, we feel the potential of our body to act. But also while acting with an object, a range of potential actions unfolds into the experience. For example, a baby grabs a toy to chew on it, but while moving it towards her mouth, she discovers sound coming out of the toy. This guides her into shaking the toy as she discovers it is not only chewable, but also shakable. Thus, the hidden affordances reveal themselves in the experience of acting in the world. These affordances emerge during interaction in a certain perceptual order, with those presented at a distance, such as the visual ones being revealed first, and the more intimate ones, such as tactile and haptic ones, surfacing into the experience during the physical manipulation. Direct interaction can also lead to disappearance of certain affordances. A dry leaf affords pulverisation, but only once, as it will not reconstruct itself. Through such action, the leaf disappears, turns into dust, which then affords new actions. Thus, through the affordance of pulverization, a leaf triggers an irreversible process that creates new material states and new affordances.

With dynamic materials such as Silly Putty, the situation is more complex as the affordance may not only be hidden, but also unstable. The affordances of dynamic materials transform together with the qualities of the material while one observes or engages with it. When applying pressure, the solid turns fluid, directly changing the potential of the material for our performance, and its performance. As it feels solid in our hands, we may want to throw it somewhere, expecting it to fly, but then while flying the material will miss the pressure of our hand, change its state into liquid and modify its path. Therefore, the material transformation produces a kind of *fluid affordance*, which changes the potential for action offered by a material. As the material's state and qualities change, so does the relationship between it and its surroundings, including us. And we are invited to explore and to accommodate these new situations as they present themselves.

These *fluid affordances* challenge the order of who affords something for whom. Most often, designers focus on how everyday objects afford human actions and how to choreograph their emergence as the experience unfolds. But what happens when we shift this other way around, and ask how a human can afford materials' actions? Human affordances have been considered only in relation to other humans: the so-called social affordances;[19] or in relation to oneself as a kind of bodily awareness or the bodily affordance.[20] But with active materials, one has to adapt oneself to materials that may not obey one's wishes. This material disobedience is both attractive and irritating. But in any case it leads us to construct seed spaces for materials to act, while we try to follow and adapt to their *fluid affordances* as they emerge.

Thus, *fluid affordances* are dynamic and invite exploration, rather than use. They also allow us to engage with important criticism of affordance as instrumental and fixed. Active materials destabilize this idea in an embodied way. As the qualities of active materials change, their related potentials and affordances raise into experience. *Fluid affordance* is an articulation of affordance, which emerges by engaging with dynamic materials. It allows us to work with and think about the instability of material transformation while in the midst of it.

Liquid Stories

Material transformation was the theme of the third workshop day. We mixed different liquids to activate their self-organizing processes. We prepared conditions for Belousov–Zhabotinsky reactions, played with home-coated photonic crystals suspended in oily fluid and did freestyle mixtures of acids, bases, oils and watery

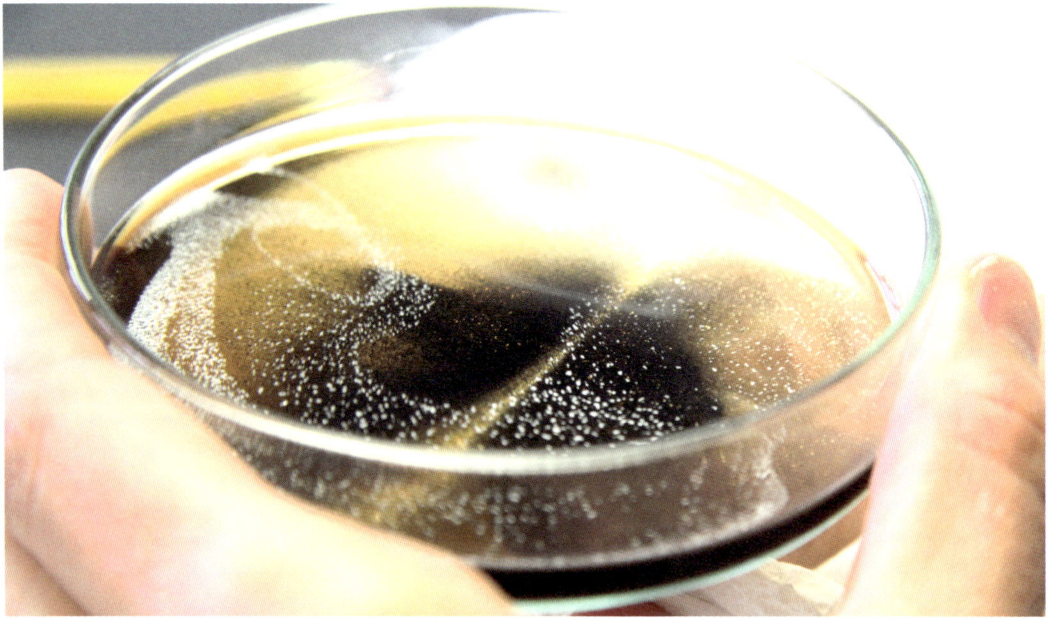

Fig. 5: Photonic crystals exposed to magnets.

solutions. As these fascinating transformations occurred in front of our eyes, we became more and more silent, seduced by material activity. However, some mental buzz was still in the air as one of the participants insisted on an account explicating what was happening. Rather than doing so, the artists who prepared the materials performance invited him to interact with it. As soon as he moved from observing to acting with matter, the participant was pulled into material process and left his questions aside for a moment.

As we became a part of the ongoing transformation, our ever-investigative minds became quieter. Materials often needed only a trigger to activate a process, such as a drop of one liquid into another. At this point, we could observe or intervene further. We tried to influence the process gently, without blocking it: blowing onto the active parts of surfaces to spread its activity; touching dry ice particles spinning on a liquid to divert their paths; swinging a magnet close to a container to activate photonic crystals changes (Fig 5). We were with transforming materials rather than observing and analyzing them from a distance.

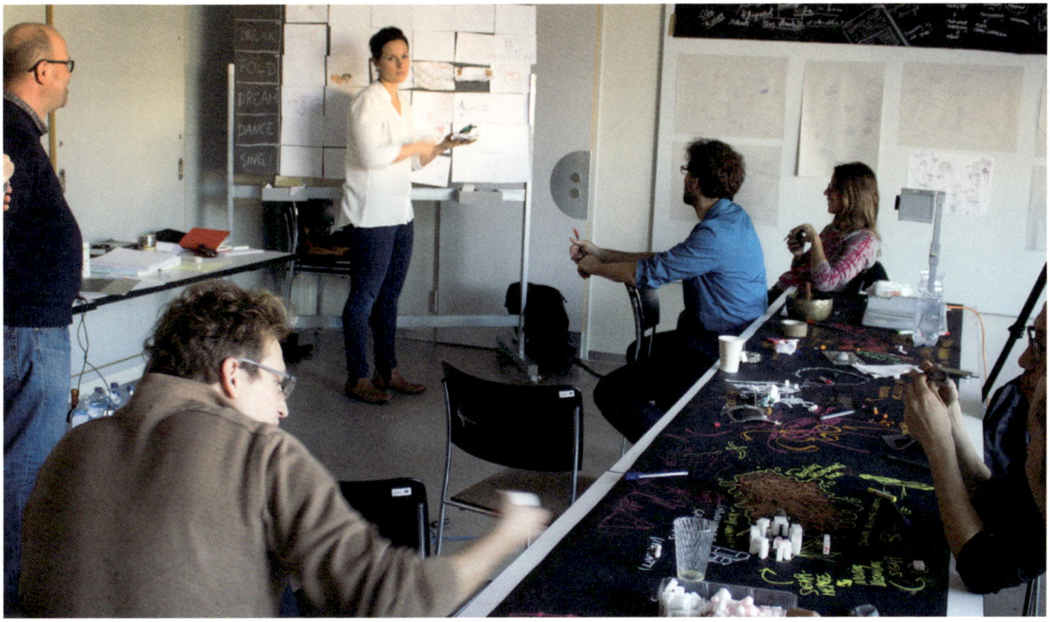

Fig. 6: Participant presenting one of the ideas resulting from the *Design Speed Dating* exercise. Long black paper covered with sketches and material arrangements.

Charged by the materials' activity and presence, we moved into searching for potential associations, situations and stories. To prepare a framework for *Design Speed Dating*,[21] together with all participants we composed a matrix of locations (volcano, tram, ocean, toilet, Mars…) and actions (contemplate, fold, dance, break, sing…). We placed a matrix on a blackboard and a blank black paper and materials on the long table. Participant couples had to chose two written dimensions: the location and the activity, in addition to the third one, given by a material physically present on the table between them (Fig 6). They then sketched or described stories involving the three dimensions on the piece of paper and put it back onto the matrix on the board. After ten minutes, participants would switch to a next combination of dimensions, a new material and a new brainstorming partner. When the matrix was fully populated, each participant presented his or her favourite story to the group.

One story involved a volcano as a location, contemplate as an action and the magnetic Silly Putty. The magnets in the story moved the magma so that the volcano could contemplate itself through moulding. The lava broke out, cooled down from dense activity to less activity and learned of its boundaries. The volcano produced a haze of abstraction, a dizzy sphere evaporating from solid stone. The magnets trained the volcano to create a crystal structure, which broke like Silly Putty but had magnetic qualities. Similarly to this one, other stories equally featured the performance of materials, with human actors rarely present. While describing these, the participants often wanted to refer to the sculptural traces of their ideas on the table. However, only the embodiment of the last idea was to be found on the table, but it carried the history of the previous modifications. In fact, the material chosen for this story, the magnetic Silly Putty, did not exist in the beginning of the exercise, but was formed during previous speed dating sessions. In that sense, the ideas travelled not only through people, but also through the materials, which changed their shape in each brainstorming session.

In *Design Speed Dating*, the use of materials showed some differences in comparison to the same exercise using objects in the workshops we conducted in the past. Firstly, the materials were manipulated by participants more than the artifacts were. This led to having the same material featured in different stories in many ways. For example, glowing sugar cubes were turned into a landscape, a movement machine and a cloud of thoughts. Secondly, the resulting stories were thus more unexpected and open ended than in the workshops using objects. Materials did not figure as passive objects at the service of humans, but were more autonomous and dynamic. Thirdly, the aesthetic dimension of the materials played a more prominent role. Compared to the *Design Speed Dating* using everyday objects, social and emotional aspects of the story were overshadowed by sensory accounts and material changes.

On the last day of the workshop we tried to reflect on our thoughts and experiences from the previous days. We diagrammed and discussed concepts and actions enabled by different materials. Among many other actions, the Silly Putty afforded the transfer of the portrait photograph of one participant onto its surface. As it was a highly malleable material, other participants started damaging the infused image by inevitably manipulating the material. The person whose image was being destroyed argued that the Silly Putty was not a good material for image support. But this again was the instrumental consideration of a material, which must serve a certain purpose, the path we tried to avoid. In fact, participants manipulating the image disagreed and felt they were preforming a rich sensory and thought experiment. For them, a new kind of image-material experience emerged through the tactile manipulation, which was very different to manipulating an image on a paper support. Rather than crumbling, folding or tearing an image, it was stretched, compressed, and melted with the Silly Putty, and incorporated with other materials. Subverted and repurposed beyond its functional affordances, we felt that the image's potential increased.

We realized that not only our everyday habits, but also our established academic approaches sometimes took us off the experimental path. Not only the image support argument, but also the concept of agency generated a lot of controversies, as it has been criticized as a superfluous concept. While many agreed we must do away with the notion of agency, most artists and designers on the team felt the notion was important for their creative process, while others with background in theory thought it was useless. Although verbally an agreement was not possible, this conflict was the productive one, as we turned to materials' capacity to act in an embodied way. Without using established terms such as agency, we could agree that actions are distributed and do not belong exclusively to the human, to the material, or to the environmental. Although active materials themselves helped us go around our habits, we felt that we needed stronger constraints that would help us to reduce mental noise cause by our established ways of (disciplinary) thinking.

Reflections…

This text provided some insights into participatory ways of thinking and creating with active materials. The *Material Aktiv Denken* event, with its techniques and resulting experiences, served as a case study as well as a trigger for articulating the concept of *fluid affordance*. The event may have opened more questions than it answered in examining what happens when we give a more active role to materials in participatory research contexts. By focusing on experience with and of active materials, we probed different workshop conditions and theory-practice techniques with the intention of merging material and conceptual processes.

Examining the role of active materials led us to reflect on the idea of affordance both in theory and practice. The affordances of materials, and particularly of active materials, proved to be different from those of an object. Working with material transformation pushed us to consider affordances in a more unstable and process-oriented way. The materials were manipulated in the workshop more then the objects were. The participants felt entitled to mold or even destroy a material assemblage as it was presented. Rather than their use and purposefulness, the aesthetic dimension of these materials was the key in creating and discovering affordances.

These differences between the status of objects and materials in participatory workshops reflect how the cognitively established affordances of everyday objects may limit imagination and thought in an instrumental way. Especially in brainstorming, participants must quickly come up with an idea, and they will most likely pick up the most obvious and habitual affordance of an object. As proposed above, active materials can help us go around this issue because they invite experimentation to begin with. Their *fluid affordances* enable open-ended processes and ideas, as exemplified by the material experiences and stories of the *Material Aktiv Denken* workshop.

Reflecting on the human roles in this workshop in respect to active materials, we can identify three perspectives. First, the organizational perspective is the one of creating and preparing the seeds for flows of materials and thoughts. The organizers must search for ways of activating and constraining materials, discussions and other human activities. In the case of liquid materials, the seeds can be enclosures and containers (e.g. the hard boundaries of a container and a soft osmotic boundary of a spherified cocktail), other material forces (e.g. an activating drop) and human actions that shape and enable their flows (e.g. blowing air). Thus, when creating seed-spaces for material events, we must consider not only materials, but also their physical surroundings and the influence of people in them. Thus, the organizers must think the event in terms of continuous materials-environments, rather than separate and stable objects and contexts.

The participant perspective can be an observational or a performative one. During observation, one experiences flow and transformation without intervening. While we hope to circumvent analytic processes through material experience, it may be challenging for participants to resist the mental buzz and comments on material transformation while it is happening. The performative perspective, which is often reserved for the artist or the researcher, forms possibly the most important part of the workshop if opened to other participants. Direct engagement with flows and transformation proves to increase the presence of participants while also destabilizing their expectations, assumptions and beliefs. Participants can discover hidden and emergent *fluid affordances*, and then choreograph them to enable or to block the flow of material activity. They are facing the active material, which is waiting to be given a possibility to transform and to show more than its potential to perform an existing predefined function. Such performative material

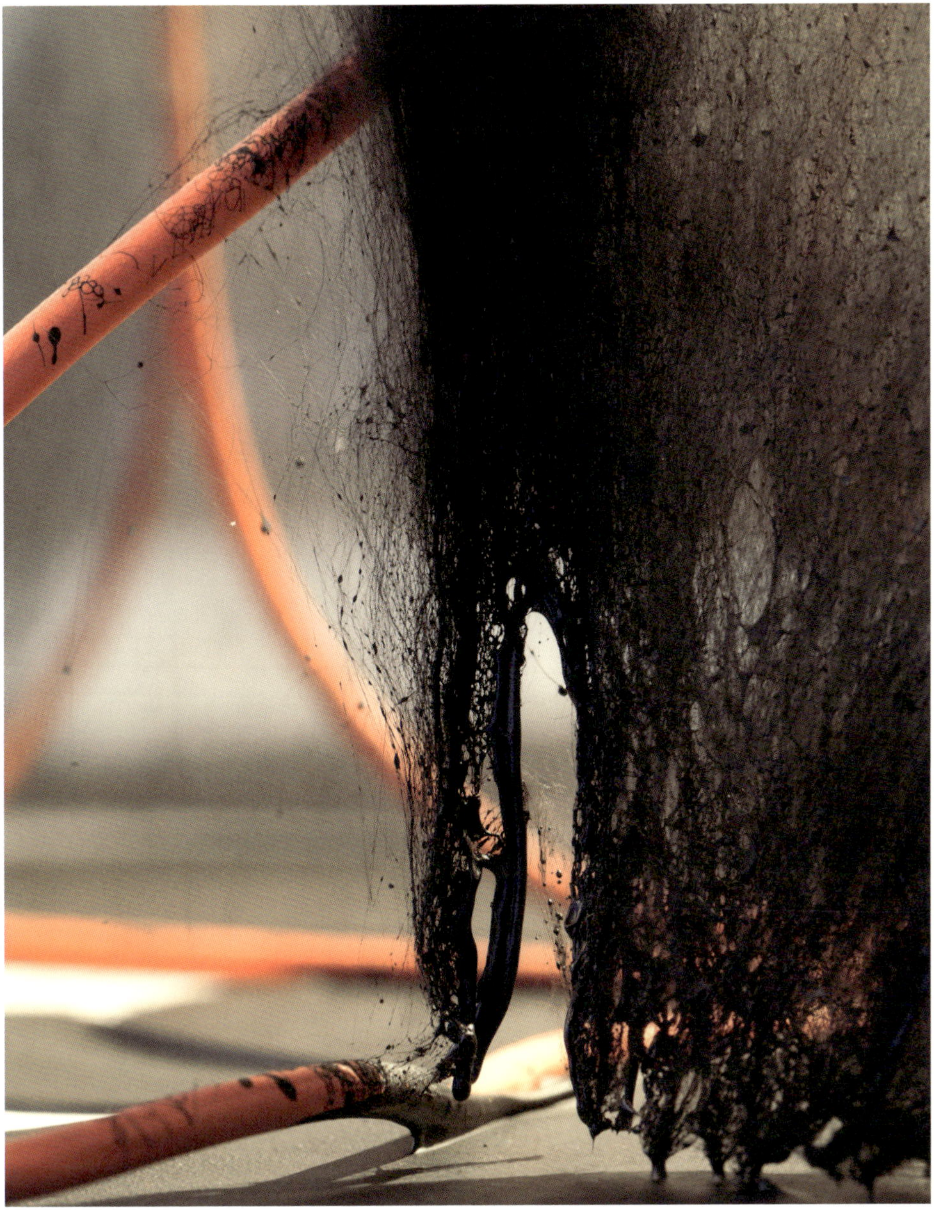

Fig. 7: Detail of Leaking (for the time being) by Roman Kirschner
as shown in the exhibition Kontinuum, gallery Im Ersten, Vienna, Feb. 2015.

experiences can directly question not only our habitual ways of doing, but also our established ways of thinking.

… and Leakings

While I could formulate some conceptual proposals stimulated by the *Material Aktiv Denken* workshop, arguing how and if artistic practices have benefited from this event proved to be more difficult. Although repercussions of the workshop into practice cannot be clearly articulated, they can be felt in the artworks coming out of the project, Liquid Things. *Leaking (for the time being)* by Roman Kirschner (2015) is a processual sculpture (fig. 7 and pp. 88-89). An upside down table of an unusual shape, hangs from the ceiling, filled with black liquid. A sudden sound is accompanied by an inflation of a bubble somewhere on the black landscape. A subtle "pufff…" is heard as the bubble bursts, leaving a tiny crater on the surface. At the same time below the table: more activity. The liquid slowly drips, forming long threads and populating the space and the white surface with sculptures and drawings.

Over several weeks, the liquid surface in the table thickens in some places, and rises in others. The landscape that has formed now sounds deeper and subtler. Below, the threads form a complex, quasi-architectural landscape shaped by dripping and the arrangement of red hanging lines, but also by the specific environmental conditions and human presence. The differences in air temperature create an airflow which slowly shapes the curvatures of this black-red landscape. The sculpture leaks into the environment, while remaining open to its dynamic influence. The hanging table with its holes, the controlled and uncontrolled airflow, the gravity, the room, the artist and the audience subtly shape this sculptural-environmental piece.

1 A comment from a participant using the *Fluid Morphologies* installation. Franzke, Luke; Rossi, Dino and Franinović, Karmen, *Fluid Morphologies: Hydroactive Polymers for Responsive Architecture in ACADIA16: Posthuman Frontiers, Proceedings of the 36th Annual Conference of the Association for Computer Aided Design,* in: *Architecture*, 2016.

2 Barad, Karen, *Posthumanist performativity: toward an understanding of how matter comes to matter,* in: *Signs: Journal of Women in Culture and Society*. 28 (3), The University of Chicago Press 2003, pp.801, retrieved Aug. 26, 2016, from: http://www.jstor.org/stable/10.1086/345321.

3 Massumi, Brian, *Semblance and Event: Activist Philosophy and the Occurrent Arts*, MIT Press 2011, pp.12.

4 See the *Enabling* page on the *Enactive Environments* website, retrieved Aug. 26, 2016, from http://www.enactiveenvironments.com/category/enabling

5 One example is our work with irregularities of electroluminescent paper, which must be perfectly homogenous in commercial products. See Franinović, Karmen and Franzke, Luke, *Luminous Matter:*

Electroluminescent Paper as an Active Material in Aesthetics of Interaction: Dynamic, Mulitsensory, Wise, in: *Proceedings of the 9th International Conference on Design and Semantics of Form and Movement*, DeSForM 2015. pp 37-47.

6 See documentation of the *Material Aktiv Denken* workshop, retrieved Aug. 26, 2016, from http://www.liquidthings.net/?page_id=1295

7 Quetglas, Peep, *Silk kite in the Wind*, Foreword to Conde, Yago, *Architecture of the Indeterminacy*, Actar 2000, p. 6.

8 Franinović, Karmen, *Enactive Encounters in the City,* in: Beesley, Philip et al. (eds.), *Responsive Architectures : Subtle Technologies*, Riverside Architectural Press 2006, pp. 74-79

9 Pask, Gordon, *The Architectural Relevance of Cybernetics,* in: *Architectural Design*, September 1969, pp. 494-496.

10 Franinović, Karmen, *Architecting Play in Journal of AI & Society - Special issue on Poetic and Speculative Architectures in Public Space*, Vol 26.2, Springer 2011, pp. 129-136.

11 See *Contraction and Expansion* (2012) workshop documentation, retrieved Aug. 26, 2016, from http://www.liquidthings.net/?page_id=1000

12 Müller, Michael, *Participatory design: the third space in HCI,* in: Jacko, Julie and Sears, Andrew (eds.), *The human-computer interaction handbook: fundamentals, evolving technologies and emerging applications*, CRC Press 2002, pp 1051-1068.

13 Franinović, Karmen, Chapter 6: *Participatory Methods: Engaging Senses in Amplifying Actions: Towards Enactive Sound Design*, Doctoral Thesis, University of Plymouth and Zurich University of the Arts, 2013, pp 159—200, retrieved Aug. 26, 2016, from http://pearl.plymouth.ac.uk/handle/10026.1/1496

14 This technique was developed in a series of seven workshops on sonic interaction, in which participants introduced themselves by imitating the object they brought with their voice (see *Voicing the Object* exercise in Franinovic, 2013, pp. 183.

15 We used this experimental techniques for the first time in the *Dancing the Virtual* event organized by Erin Manning and Brian Massumi in Montreal in 2005.

16 Transcript of the video recording from the workshop.

17 Norman, Donald, *The Psychology of Everyday Things*, Basic Books 1988, p. 9.

18 Gibson, J. James, *The Ecological Approach to Visual Perception*, Houghton Mifflin 1979, p. 127.

19 Loveland, Katherine, *Social Affordances and Interaction II: Autism and the Affordances of the Human Environment,* in: *Ecological Psychology Journal*, Vol. 3 (2), 1991.

20 de Vignemont, Frederique, *Bodily Affordances and Bodily Experiences,* in: Coello, Yann and Fischer, Martin (eds.), *Perceptual and Emotional Embodiment: Foundations of Embodied Cognition*, Volume 1, Routledge 2016.

21 The first version of the *Design Speed Dating* exercise used the matrix dimensions described by words only (see *Urban Atmospheres* workshop, retrieved Aug. 26, 2016, from http://www.urban-atmospheres.net/Ubicomp2005/). Substituting one verbal dimension of the matrix with an embodied dimension proved to facilitate idea generation in a transdisciplinary setting (Franinovic, 2013).

Sources of the figures:

Fig. 1: Photo courtesy of The Phillips Collection. Photo credit: Alexander Morozov.
Fig. 7: Photo credit: Roman Kirschner

Acknowledgements

Thanks to all the participants of the workshop, especially to Roman Kirschner for co-organizing the workshop and introducing me to Silly Putty and to Clemens Winkler for helping with the organization and for chocolate clouds.

Biographies

JEAN-MARC CHOMAZ is a physicist and artist. His scientific research encompasses areas such as the dynamics of soap films, global instability, vortex breakdown, geophysical and stratified fluids, and biomechanics. He is a director of research, professor and chair at the University of Paris-Saclay, where he co-founded the Hydrodynamics Laboratory he co-directed from 1990 to 2013. With more than 25 different artworks, performances and installations, his art practice aims at exploring the sensory and emotional foundations of science.

EVELINA DOMNITCH AND DMITRY GELFAND create sensory immersion environments that merge physics, chemistry and computer science with uncanny philosophical practices. Having dismissed the use of recording and fixative media, their installations exist as ever-transforming phenomena offered for observation. In order to engage such ephemeral processes, the duo has collaborated with numerous scientific research facilities.

KARMEN FRANINOVIĆ is professor for interaction design at Zurich University of the Arts. Her research focuses on enactive approaches to cognition and design, responsive environments, sonic interaction, creative methods and active materials. Karmen founded Zero-Th studio and Enactive Environments Lab, whose work has been shown internationally at venues such as MoMA Ljubljana, Cambridge Junction, and Centre Pompidou.

INGE HINTERWALDNER is visiting professor for modern and contemporary history of art and imagery at Humboldt University in Berlin. Her current research interests are the relationships between the arts, sciences and technology, model theory and operativity of images, temporality in the visual arts, computer-based art and architecture.

ROMAN KIRSCHNER works in the fields of plastic arts and art-based research. From 2012 until 2016 he led the project Liquid Things – Research on Active and Transitive Materials at the University of Applied Arts Vienna. His art projects have been shown in numerous exhibitions in the USA, South America, Europe and Asia. In his current research he is developing a paradigm of material activity in the plastic arts.

ESTHER MOÑIVAS is professor for art history and aesthetics at Nebrija University, Madrid. She holds a doctorate from Complutense University and is currently leading the Nebrija Group of cross-sectional surveys in contemporary artistic creation (ETCC). Her ongoing research interests are new media and materials in current artistic creation, intersecting creative processes between art and science, and the aesthetics of fluid materials.

HANS-JÖRG RHEINBERGER is honorary professor for the history of science at the Technische Universität Berlin and holds an honorary doctorate from ETH Zürich. Formerly director of the Max Planck Institute for the History of Science in Berlin, he is a member of the Berlin-Brandenburgische Akademie der Wissenschaften and the Leopoldina. His current research interests are the history and epistemology of experimentation and the relationship between the sciences and the arts.

BENJAMIN STEININGER is a cultural and media theorist, historian of science and technology, and curator in Vienna. He has written a book on the history of the German motorway system and a dissertation on the industrial history of chemical catalysis in the 20th century. His main research fields are the cultural history of energy and acceleration and the history and theory of the materials of modernity: fuels, fossil raw materials and building materials.

Imprint

ROMAN KIRSCHNER, Project LIQUID THINGS,
University of Applied Arts Vienna, Austria

Library of Congress Cataloging-in-Publication data. A CIP catalog record for this book has been applied for at the Library of Congress.

Bibliographic information published by the German National Library
The German National Library lists this publication in the Deutsche Nationalbibliografie; detailed bibliographic data are available on the Internet at http://dnb.dnb.de.

ENGLISH EDITING: Christopher Barber, Darcy Alexander and Scribendi
COVER PICTURE: Roman Kirschner
GRAPHICDESIGN: Theresa Hattinger www.thehatdesign.com
PRINTING: Holzhausen Druck GmbH, Wolkersdorf, Austria

Printed on acid-free paper produced from chlorine-free pulp. TCF ∞
Printed in Austria

ISSN 1866-248X
ISBN 978-3-11-052395-9
This publication is also available as an e-book (ISBN PDF 978-3-11-052599-1)

www.degruyter.com